Contents

Tables

POLICY STUDIES IN EMPLOYMENT AND WELFARE NUMBER 23

General Editor: Sar A. Levitan

SEX, AGE, AND WORK

The Changing Composition
of the Labor Force

Juanita Kreps and
Robert Clark

The Johns Hopkins University Press, Baltimore and London

This study was prepared under a contract to the Task Force from the Office of Research and Development, Manpower Administration, U.S. Dept. of Labor.

Manufactured in the United States of America

The Johns Hopkins University Press, Baltimore, Maryland 21218
The Johns Hopkins University Press Ltd., London

Library of Congress Catalog Card Number 75-34452
ISBN 0-8018-1806-0 (cloth)
ISBN 0-8018-1807-9 (paper)

Library of Congress Cataloging in Publication data will be found on the last printed page of this book.

Figures

Preface

In the 1970s the nation's economic policies are concerned with the major problems of unemployment and inflation, the latter exaggerated by limitations on the supply of oil and a rapid escalation in the price of all energy resources. Expectations of a quick end to the acute recession that developed midway through the decade proved far too optimistic, but business forecasters—a resilient lot—continued to shift the turnaround date, quarter by quarter, into the future. The years of 1974 and 1975 were years of poor economic performance when measured by any one of a number of indices, the most frequently cited being that of unemployment.

In their attempts to explain those events, some economists turned increasing attention to the inflation-unemployment trade-off, following earlier studies of the Phillips Curve and a subsequent movement of the curve to the right, indicating both higher unemployment and faster rates of price rise. Others rejected the notion of any interconnection between price and unemployment changes and argued instead that controls on spending and the money supply would not only contain inflation, but would also reduce unemployment. One of the federal policy manifestations of the latter view was an attempt to limit federal expenditures, despite a public clamor for job creation.

Because of the inflationary threat the best-known tools for combating unemployment were thus temporarily shelved, despite the highest rates of idleness in nearly half a century. Although price increases seemed to be slowing during the latter part of 1975, there was little hope that the price pressure would abate sufficiently to allow the fiscal freedom envisioned by early Keynesian advocates. Instead, policies aimed at mitigating the unemployment problem in the short run—extended unemployment compensation, the creation of public service jobs, special programs for youth, for example—received increased attention.

Factors associated with the demand for labor have been stressed in the economic literature during the first half of the 1970s, as the problem of job shortages became more acute. Although the treatment of demand factors has not been exhaustive, nevertheless it is important to keep in mind some changes that are occurring on the supply side, lest longer-range shifts in labor force composition create still other structural impediments to full employment. These changes in supply may call for different kinds of policies and programs; they may be contributing to or concealing some portion of those classified as unemployed; they may signal the need for a major re-examination of the content and characteristics of the labor force. Specifically, shifts in the sex, age, and marital status of the members of the work force, along with the broader question of the allocation of time over its three major uses—market work, home work, and leisure—call for further study.

The tools for such study are available, and much of the theoretical framework has been developed. Gary Becker's analysis of time allocation and his subsequent article on the theory of marriage, and several studies that his work has inspired, provide a basis for researching many questions formerly thought to lie outside the economist's purview. The literature on investments in human capital has also grown rapidly in recent years. New sources of data, particularly those now available from Herbert Parnes' surveys, make it possible to draw comparisons between the labor force behavior of different age and sex groups, particularly workers at the beginning and near the end of worklife.

Formulation and analysis of the critical questions having to do with the allocation of time between work (both in the market and in the home) and nonwork pursuits and the distribution of work over the age groups and between the sexes are required. For although each of us is given the same twenty-four hours per day and, with relatively few exceptions, each of us lives to old age, the amount of time we spend at work varies widely. Within a family, working time differs within an even broader range, depending on whether one or two adults enter the labor force, on the number of children and hence the amount of home work required, on the stage of the family's life cycle, and on the taste for leisure. What do we need to know about the way working and nonworking time has come to be apportioned as we enter the last quarter of the twentieth century? And what do we anticipate will happen to the present allocation of time as we move toward that magical year 2000? The policy implications of major shifts in the distribution of market work between men and women, and between age groups, need careful review.

To identify the questions, it is useful to examine the changes that have occurred in recent decades in the age and sex distribution of market work and to ask whether similar reallocations of work have taken place in the home as well. In the review provided here we have concentrated on the empirical record, only occasionally fitting events into the framework of recent models. Our purpose has been to focus primary attention on the way in which shifts in the apportionment of market work are correlated with other changes: reallocation of home work; a possible redistribution of income between income classes and age groups; most fundamental of all, a transition in life styles that marks a significant departure from traditional family patterns.

As the composition of the labor force undergoes important changes, workers' expectations change as well. The employee late in the twentieth century, who more often will be female, more often not married, more often a college graduate, is likely to have a view of work that is quite different from that of the blue-collar male family head. The characteristics of the new workers, and their aspirations in particular, may call for some drastically different manpower policies in the future.

J. K.
R.C.

Durham, N.C.

Sex, Age,
and Work

1

The Allocation of Work and Free Time

During the last half-century there have been important changes in the work roles of young and older men and of women at all ages. Among men, extended schooling has postponed entrance to the work force by several years; meanwhile, higher job requirements have made it difficult for youth to find employment without post-secondary education. At the other end of the age spectrum, men have come to expect retirement at age 65 as the outside limit, with more than half of those retiring in recent years leaving their jobs before reaching that age.

With women the changes have been far more dramatic and have touched all age groups. Beginning with women's assumption of wartime jobs in the early 1940s, their numbers in the labor force have continued a sharp climb. Now, three and a half decades later, their work profiles reveal much steadier commitment to market work, with uninterrupted labor force participation through the childbearing and early child-rearing period rapidly becoming the norm, particularly for educated women. If educational levels continue to rise and fertility continues to decline, the worklife pattern for married women will come to resemble more closely than previously those of men and single women.

The cumulative effect of these changes in women's worklives has not been fully recognized in the literature nor integrated into

1

public policy. Their continued participation in the labor force following World War II; a subsequent resumption of the long-run decline in birth rates; increasing interest in lifetime careers; a rapid growth in the proportion of women who are single or divorced—all these factors interact in a manner that suggests a basic and permanent change in the attitudes and expectations of both sexes toward work roles. The shifts in work are as significant in the home as in the marketplace, although the latter has received far more attention.

As women have moved from the home to the marketplace, and men have reduced the portion of their adult lives spent on the job, the division of time between work and leisure also has shifted. Aggregate data revealing somewhat shortened working hours per week, added time free for vacations and holidays, and a significant increase in free time for males before and after worklife obscure individual variations in males' working patterns. In particular, the longer hours worked annually by men in certain types of jobs[1] and the extension of worklife beyond the usual retirement age reflect strong work commitments and high levels of market demands for the services of this group, along with a limited supply of persons available to meet the demand. In general, such positions require extensive training and continuous work experience, as well as access to training programs. Since these characteristics have been scarce among women in the past, there has been relatively little erosion of the males' dominance in the higher-level professions.

In other types of jobs, however, for which only high school or even college-level training is required, the male's time free of work has grown substantially during the last half-century,[2] as economic growth has conferred on workers both higher real incomes and more free time to enjoy them. Insofar as the man's increase in free time was offset by an increase in market work done by his wife, growth in the family's leisure may have seemed illusory. But was it? Have changes in the nature of household tasks tended to offset the woman's increased market work, declining sufficiently, perhaps, to match the time she now spends in the labor force? Have other family members, including the husband, shared the home duties? Or has the married woman's time on the job simply increased her total workload, even as that of her husband declined?

Market work, while becoming more freely shared between the sexes, appears not to have been paralleled by similar redivisions in home work, although the evidence is not at all clear on changes in the distribution of work within the household. A 1971 study concluded, "Recent research has indicated a trend toward a division of labour in households which is less 'segregated' along traditional stereotyped lines and more 'interchangeable' or 'joint'— i.e., shared by husband and wife."[3] But a slightly earlier survey of husband-wife families indicated that working mothers were likely to shift home work to teen-age children but not to husbands; indeed, husbands averaged 1.6 hours daily on home work whether or not their wives had market jobs.[4] Family work roles also have changed very little. Wives have continued doing most of the work inside the home and husbands the maintenance and yard work.[5] Looking toward the future, Estelle James speculates on the effect of woman's gaining wage parity with man: will her increase in relative earnings not bring a shift in home responsibilities? Her lower earnings in the past made her time in the market less valuable than his, and she compensated by doing the home work. When the wage difference no longer holds, the household division of tasks seems likely to change.[6]

The male's utilization of time freed of market work in the performance of daily household chores is hampered both by the form in which much of the new leisure appears and by other work-related pressures, such as commuting time. The division of the male's free time between workyear and worklife reductions has favored the latter heavily, and as a result men have gained long periods of relative freedom from work at those times when households least need their assistance—before marriage and after the family has shrunk to two persons. Were the nonworking time apportioned more evenly over the worklife, it would greatly enhance the male's availability for home work at critical times in the family's life cycle.

A future reordering of working time could entail greater flexi-

3

bility in the spacing of years of work and years of nonmarket activity. More specifically, continuous years of work would seem to produce a growing disutility for an additional year of work, while extra years of retirement may hang heavily over the individual in later life. If at age 20 one were given 45 years of work and 10 years of nonwork time, and then allowed to allocate his time as he wished, he probably would not elect to work straight through to age 65 and then retire. Currently, government and industry regulations bias a work choice toward this allocation, despite the fact that the maximization of lifetime utility may call for a different pattern of time use. Removal of the constraints would allow greater flexibility and should increase lifetime utility.

Current patterns of allocating time between market work, home work, and leisure result in part from women's greatly increased labor force participation; in part from temporal shifts that tend to crowd work, more and more, into the middle years; in part from cyclical downturns that have reduced working time and produced differential rates of unemployment by age and sex; and finally, from intrafamily decisions that divide home work, presumably on the basis of skills, preference, and power. Each of these components of the change has been studied separately, in pursuit of answers to different questions.

But the total impact of the redistribution of market and nonmarket work between sexes and age groups has not been examined, although the implications of the emerging work and leisure-time patterns raise important questions. Consider, for example, the possible effect of the new divisions of work on investments in human capital. Will the amount spent on educating and training women grow to equal that of men, as returns to women's work grow to equal those of men? How will the distribution of earnings over the life cycle be affected, and how will the increasing proportion of two-worker families affect the distribution of income between income classes? Will unemployment increase as women, who have higher rates, join the work force? Or will their greater attachment to the labor force reduce their unemployment? Perhaps most important, will the newer distributions of work offer workers wider options as to the range of possible jobs, as well as greater freedom of choice as to when and how to take their free time? If so, the quality of life could be improved greatly. In the ensuing analysis, the first step is to review the worklife changes for married women and men, taking particular note of the relationship of human capital investment to the allocation of household time. But since there is now a trend toward delaying marriage and since divorce rates continue to rise, it is clear that a growing proportion of the work force will come from adults who are not married. The work activity rates for nonmarried adults, male and female, appear to be different from those of their married counterparts: nonmarried women have higher and nonmarried men lower participation rates. On the assumption that these differences continue to hold, the implications for the labor force of a growth in one-adult families are indicated. The concept of a lifetime utility function is then developed to help explain the allocation of each family member's time over the life span of the household.

The concluding chapter returns to the questions posed above, which can be summarized under the major query: what are the implications of recent changes in the allocation of market work, home work, and free time between the sexes and between age groups?

2

Worklife Changes of Married Men and Women

The twentieth century has seen a dramatic reordering of work and free time over the life cycle. In the typical American family, the wife has come to spend a larger and the husband a smaller proportion of life in market work. While males have gradually lengthened their period of schooling and their years in retirement, females have extended their worklives by staying in the work force through the childbearing period or returning shortly thereafter. Time spent in home work has declined with the decrease in family size and improvement in household technology. Reductions in the length of the workyear, resulting from longer vacations, paid holidays, and shorter workweeks, have made it possible for the labor force participant to lengthen the time he spends on nonwork pursuits.

But growth in the potential time free of market work appears not to have brought with it the leisure and serenity some critics expected. Steffan Lindner opened *The Harried Leisure Class* with a reference to the hectic pace of modern life, lamenting,

It used to be assumed that, as the general welfare increased, people would become successively less interested in further rises in income. And yet in practice a still higher economic growth rate has become the overriding goal of economic policy in rich countries, and the goal also of our private efforts and attitudes.[1]

7

In addition to Lindner's question of whether free time actually has grown in recent decades, important issues arise in the wake of shifts in the allocation of work within the family. Change in the manner in which a family apportions the time of its members over market work, home work, and leisure has important implications for the size and composition of the labor force; the economy's productivity and rate of growth; and the distribution of income among families. The major source of such reallocations can be seen from a review of the working patterns of women, particularly those who are married.

Women's New Worklife Patterns

The labor force participation rates of married women more than doubled from 1900 to 1940 and then almost tripled between 1940 and 1970. Table 2.1 shows the sharp increases in these rates, by age groups, during recent decades. In contrast, the levels of labor force participation of single women and men have been falling during most of this century. The combination of these two trends has resulted in an increase in the portion of the labor force made up of married women. Between 1940 and 1960 the increase in this group accounted for 56 percent of the 14.4 million total increase in the labor force.[2] By 1974, the category of "married women,

Table 2.1. **Labor Force Participation Rates of Married Women, Husband Present, Age 14 and over, 1940 to 1974**

Year	Total	Under 25	25–34	35–44	45–54	55–64	65+
1974	43*		46.1	50.1	49.6	34.9	6.7
1970	40	45	38	46	48	35	8
1960	31	30	28	37	39	25	7
1950	22	25	22	27	23	13	5
1940	14	16	18	15	11	7	3

SOURCE: U.S. Bureau of Census, *Census of Population: 1970,* Subject Reports PC(2)-6A, "Employment Status and Work Experience," Table 5, p. 69. 1974 data from U.S. Department of Labor, *Manpower Report of the President: 1975* (Washington, D.C.: U.S. Government Printing Office, 1975), p. 252.

* Includes only women 16 years and older.

husband present" had come to comprise 22.5 percent of the total labor force, up from 18.6 percent in 1960.[3] Disaggregation of these trends helps to explain the continuing life-cycle changes in the family's allocation of time.

Drawing on the data in table 2.1, cross-sectional patterns of women's labor force activity are shown in figure 2.1. Prior to

Figure 2.1. Labor Force Participation Rates of Married Women, Husband Present, Age 14 and over

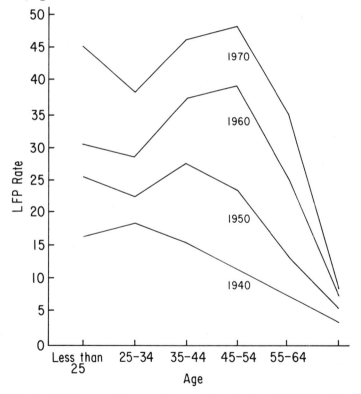

SOURCE: U.S. Bureau of Census, *Census of Population: 1970*, Subject Reports PC(2)-6A, "Employment Status and Work Experience," Table 5, p. 69.

World War II, married females entered the labor force in moderate numbers during their early years of marriage, later withdrawing from market jobs for childbearing. Those women who left the labor market did not return even after children were in school or had left the home permanently. But during the war women of all ages were called upon to fill jobs that traditionally had gone only to males. Responding to the nation's demand for greater output at a time when there was a contraction in the male work force, women were drawn into market work which offered far better job opportunities and higher wages than they had ever known. These wartime conditions also contributed to a widespread public acceptance of working wives.

The impact of these and other shifts in the supply of and demand for female workers, which produced the large upward movement of married women's participation patterns, has been well documented.[4] One important aspect of this shift has been a change in the age pattern of participation. No longer do participation rates decline continuously after age 30. Rather, married women in their thirties and older, often with school-age children, are in the labor force, either because they have returned after a brief absence or because they never left. The new pattern of activity has produced the familiar "M" cross-sectional curve of the past 25 years. The second or middle-aged peak has in some years marked the period of greatest labor force activity of married women.

But this approach to life-cycle changes, using cross-sectional comparisons, can be misleading. The lower participation rates of women aged 55 to 64 reflects a lower labor force activity throughout the lifetime of this cohort. The older the cohort of women, the smaller their investment in market-related human capital, the larger their families, the more rural their setting, and the less permanent their attachment to the labor force. A cross-sectional picture tells us little about the labor force behavior of a particular cohort of women through its lifetime. To analyze life-cycle work patterns, it is, of course, necessary to follow a cohort through its worklife. As table 2.1 and figure 2.2 show, the participation rate for each cohort rises continuously (with the exception of one ob-

Figure 2.2. Labor Force Participation Rates of Selected Cohorts of Females through Their Worklives, Ending in 1970*

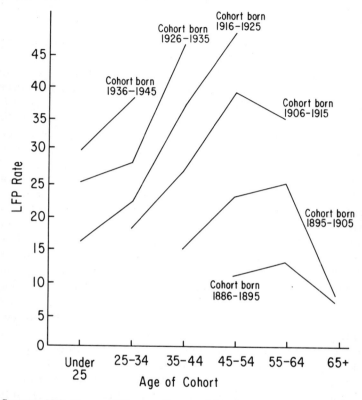

Source: U.S. Bureau of Census, *Census of Population: 1970*, Subject Reports PC(2)-6A, "Employment Status and Work Experience," Table 5, p. 69.
* The last observation for the cohort born between 1886 and 1895 is from 1960.

servation) until retirement. Within an age group, there is no evidence of a decline in participation rates during the childbearing years, as indicated by the cross-sectional patterns.[5] Moreover, at the other end of worklife, Figure 2.2 shows a tendency for the participation rate for each cohort not to decline in the forties or early fifties as indicated by the cross-sectional data. For the cohorts born 1916 to 1925, 1926 to 1934, and 1935 to 1944, the participation rate has continued to increase for as long as we have data, that is, through 1970.

Patterns of lifetime labor supply for married women derived from cohort analysis are thus dramatically different from the one drawn cross-sectionally. In addition, a higher percentage of women in each succeeding cohort has been in the labor force; that is, the lifetime pattern of each cohort lies above that of its predecessor. The cross-section M pattern, suggesting that the participation rate of married women exhibits a cyclical pattern through their lifetime, fails to describe the pattern of any cohort in the labor force.

Patterns of market activity by race show that historically nonwhite married women have participated in the labor force to a greater extent than white females. Since 1950 the differential in participation rates between all married women 14 and older and nonwhite women has remained approximately 10 percentage points, with both rates showing a 20-point increase between 1950 and 1970.[6] In March 1974, the Department of Labor reported labor force participation rates of 52.1 percent for Negro and other races and 43 percent for all married women, husband present.[7]

These rates are in large part a reflection of the different economic positions of the two groups. From the 1960 census data, Bowen and Finegan concluded that "half this differential disappears when we adjust for the effects of the other variables . . . age and schooling of the wife, children, employment status of the husband, and other family income."[8] The major explanation for the difference in participation seems to be the difference in other family income; husbands of white women had incomes substantially higher than those of black women.

Declining Labor Force Rates for Men

In direct contrast to the rapidly increasing labor market activity of married women, the overall participation rate for married men has been declining. Table 2.2 shows that the work rate of married

Table 2.2. Labor Force Participation Rates* of Married Men, Wife Present, by Age, Selected Years, April 1947 to March 1973

Year	Total, 16 and over	Under 20	20–24	25–34	35–44	45–64 Total	45–54	55–64	65 and over
April 1947	92.6	‡	‡	97.7	98.8	95.0	‡	‡	54.5
April 1949	92.2	‡	94.9	97.7	98.7	94.3	‡	‡	51.9
April 1951	91.7	96.7	95.6	98.2	98.4	93.5	‡	‡	50.9
April 1953	91.5	100.0	96.1	98.7	98.8	94.9	97.6	91.0	46.2
April 1955	90.7	98.8	94.5	98.8	98.8	93.8	97.4	88.8	44.2
March 1957	90.6	97.9	95.9	98.7	98.7	94.4	97.6	90.1	42.4
March 1959	89.6	95.7	95.6	98.6	98.9	94.0	97.3	89.3	38.2
March 1961	89.3	98.3	97.4	99.0	98.6	93.7	97.0	89.1	37.6
March 1963	88.1	97.8	96.5	98.6	98.9	93.6	97.3	88.4	32.3
March 1965	87.7	94.3	96.6	98.5	98.2	92.8	96.8	87.1	31.1
March 1967†	87.0	93.8	96.6	98.5	98.2	92.1	96.6	86.0	28.8
March 1969	86.9	95.6	95.0	98.3	98.2	91.6	95.9	86.0	30.9
March 1971	85.9	90.9	94.8	97.8	97.9	91.2	96.0	85.1	27.8
March 1972	85.5	93.5	95.2	98.0	97.9	90.6	95.3	84.5	26.6
March 1973	84.8	96.4	95.1	97.4	97.5	89.0	94.8	81.5	26.0
March 1974	83.9	93.5	95.4	97.6	97.6	88.5	94.2	81.1	24.1

SOURCE: U.S. Department of Labor, *Manpower Report of the President: 1975* (Washington, D.C.: U.S. Government Printing Office, 1975), Table B-2, p. 252.

* Labor force as percent of population.

† Prior to 1967, the labor force included persons 14 and 15 years old. In 1967 and later, only persons 16 years and over are included.

‡ Not available.

men over age 16, wife present, has fallen from 92.6 percent in 1947 to 83.9 percent in 1974. Participation for men aged 20 to 54 has remained a relatively stable proportion of the noninstitutionalized civilian population of that age group. The systematic decline in male population rates is due therefore to the reduction of labor force activity of older and younger men.[9] Since only one male in ten under the age of 20 is married, we will discuss shifts in patterns of participation of young men in Chapter 3.

The decline in market activity has been most evident in the 65-and-over age group. The LFP rate for these men dropped from 54.5 percent in 1947 to 24.1 percent in 1974, with most of the decline occurring before 1965. There also has been a steady but less steep drop in the rates of men aged 55 to 64, whose participation rate has fallen 10 points in the last 20 years. Since the late 1960s there has been a slight decline in the LFP rates of married men aged 45 to 54. Early retirement, both voluntary and involuntary, plus a relative disadvantage in the amount invested in human capital combined with higher retirement benefits, have stimulated the fall in labor market activity of older men.[10]

Additionally, the rise in health benefits, both public and private, has allowed older men to retire with disability pensions, whereas they previously would have been forced to remain in the labor force. The Parnes data on white males aged 45 to 59 show that for men who reported no health problems there was a decline of 1 percentage point in labor force participation between 1966 and 1969. However, the work rate of males who developed health problems during this period dropped by 16 percentage points.[11]

In summary, the market activity of married men aged 20 to 54 remains at high levels. However, males are leaving the labor force for retirement at younger ages, while the age of entry to market jobs has risen. To show these patterns, LFP rates are drawn for 1953, 1963, and 1973 (figure 2.3). The cross-sectional patterns for the three periods almost coincide for those men aged 20 to 54, while the older groups have lower participation rates each successive decade. Because of the stability of the cross-sectional patterns, a cohort analysis of males born between 1920 and

Figure 2.3. Labor Force Participation Rates of Married Men by Age, 1953, 1963, and 1973

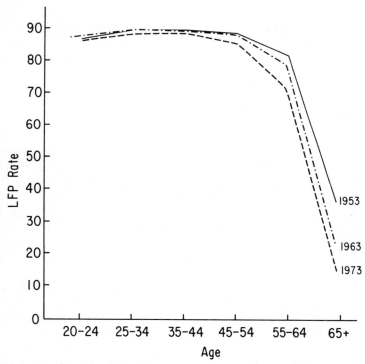

SOURCE: U.S. Department of Labor, *Manpower Report of the President: 1975* (Washington, D.C.: U.S. Government Printing Office, 1975), Table B-2, p. 252.

1933 would closely resemble the cross-sectional trend. The cohort approach would, however, show steeper declines in LFP rates at retirement than do the cross-sectional patterns.

Racial differences in the work capacity of prime-age males have been of considerable interest in recent years. Census data indicate that the difference in participation rates between all married men, spouse present, aged 25 to 54, and nonwhite males has been between 2.4 and 2.9 percentage points for 1950, 1960, and 1970.

15

A greater difference—approximately 5 percentage points—is observed for the two groups of males aged 55 to 64.[12] In March 1974, the racial difference for men aged 16 and older was less than half a percentage point—83.9 percent for all married men and 83.5 for Negro and other races.[13] Adjusting for schooling, age, and other income removes much of the observed gap in participation patterns.[14]

The Impact of Children on Time Allocation

Child care, the major service rendered in the home, has in the past required a relatively large input of the mother's time. Hence, the presence of small children has had a dampening effect on the market activity of mothers. Table 2.3 indicates that in 1974 the participation rate of married women with children under 6 was 34.4 percent, while that of wives with no children under age 18 was 43 percent. Among women there is a wide difference between the levels of market activity of those with children under age 3 (who have an LFP rate of 31 percent) and those with children 6 to 17 years (whose rate is 51.2 percent).

The impact of small children on the labor force activity of their mothers is shown in the longitudinal survey data gathered by Parnes and his associates. In their sample, white women aged 30 to 44 who had no children under the age of 6 in 1967 but with at least one child under 6 in 1969 showed a reduction of 11 percentage points in their labor force participation rate. Black and white women who had one or more children under 6 in the first survey but no children under 6 in 1969 had an increase in their market activity rates of 7 percentage points.[15]

Young children in the home lower the amount of market work by the mother. For example, of the married women, husband present, who worked in 1969 and who had children under 6 years of age, only 26.6 percent worked 50 to 52 weeks; 44.6 percent worked less than 26 weeks. In contrast, 45 percent of those women whose youngest child was between 6 and 17 worked 50 to 52 weeks and only 25 percent worked 26 weeks or less.[16] In addi-

16

Table 2.3. **Labor Force Participation Rates of Married Women, Husband Present, by Presence and Age of Children, March 1960 to 1974**

Year	All wives	No children under 18	With children under 18		Under 6		
			Total	6–17 only	Total	3–5, none under 3	Under 3
1960	30.5	34.7	27.6	39.0	18.6	25.1	15.3
1961	32.7	37.3	29.6	41.7	20.0	25.5	17.0
1962	32.7	36.1	30.3	41.8	21.3	27.2	18.2
1963	33.7	37.4	31.2	41.5	22.5	28.5	19.4
1964	34.4	37.8	32.0	43.0	22.7	26.7	20.5
1965	34.7	38.3	32.2	42.7	23.3	29.2	20.0
1966	35.4	38.4	33.2	43.7	24.2	29.1	21.2
1967	36.8	38.9	35.3	45.0	26.5	31.7	23.3
1968	38.3	40.1	36.9	46.9	27.6	34.0	23.4
1969	39.6	41.0	38.6	48.6	28.5	34.7	24.2
1970	40.8	42.2	39.7	49.2	30.3	37.0	25.8
1971	40.8	42.1	39.7	49.4	29.6	36.1	25.7
1972	41.5	42.7	40.5	50.2	30.1	36.1	26.9
1973	42.2	42.8	41.7	50.1	32.7	38.3	29.4
1974	43.0	43.0	43.1	51.2	34.4	39.1	31.0

SOURCE: Howard Hayghe, "Marital and Family Characteristics of the Labor Force in March 1973," *Monthly Labor Review*, 97 (April 1974), 24; 1974 data from Hayghe, "Marital and Family Characteristics of Workers, March 1974," *Monthly Labor Review*, 98 (January 1975), 61.

tion, the number of hours worked per week is lower for mothers of small children. Twelve percent of the women with children under 6 who were at work were on the job less than 15 hours per week, while 61.5 percent worked over 35 hours. For all working women with husband present, the corresponding figures were 7.9 percent and 68 percent.[17]

The presence of children has a two-staged impact on the level of market work of married women. When the child is under 6 and the family needs to provide almost continuous supervision, the responsibility usually falls to the mother; as a result, she curtails her market activity. But once children are age 6, the school system begins to provide virtually free child care for most of the day. At this stage, the mother is "freed" from daily babysitting duties,

17

and the family can reconsider the allocation of her time. The inducement of additional income may entice the mother back into the labor force, giving her a dual career thereafter. As the child matures, he or she is able to help in performing household tasks; the time of older children thus can be substituted for the mother's time in home activities.

This explanation of female time allocation conforms to the traditional cross-sectional and cohort analysis. However, there has been a dramatic increase in the participation rate of women with children of all ages. The rate for women with children under 3 years of age has more than doubled since 1960; moreover, there is now little difference between the rates for women with children aged 3 to 5 years and those with no children under 18 years. The increase in the participation rate of married women with children has outstripped the rise in the rates for all wives and for wives with no children under age 18.

Why have women with small children increased their market activity relatively more than other married women? Several factors seem to have had an effect. First, there has been some growth in day care programs that have lowered the price of child care or made it more accessible. Second, public attitude toward working mothers has changed significantly. Finally, as successive cohorts of women become better educated and more attached to the labor market, their costs of remaining out of work have increased. These costs include depreciation on human capital and the loss of earning years, which lowers women's lifetime earnings. Changes in women's working patterns during the past two decades, particularly, reflect important shifts in their values, aspirations, and life-style preferences.

A continuation of these trends could mean that in the near future there will be little difference in the labor market patterns of women with small children and those of all married women. In 1960 the participation rate of women with children under 3 years was 44 percent of the rate of married women with no children under the age of 18. In 1974, however, the rate of women with young children was 72 percent of that of the group with no children under the age of 18. If the relative rates of increase in participation continue, career-minded mothers of the late 1970s may take only short maternity leaves and then return to the labor force.

In addition, the recent decline in fertility has had a significant impact on the participation rates of married women as a group. Since there are relatively fewer mothers of young children when low fertility conditions prevail, fewer married women are deterred from market activity. Between 1960 and 1970, there was an increase of 4.2 million in the number of married women. However, due to the prevailing low fertility conditions, the number of married women without children under 6 years of age rose by 5.4 million.[18] This demographic effect may have an even greater impact during the remainder of the twentieth century, as the United States approaches a zero population growth rate.

The presence of children appears to have little effect on the participation rate of married men between the ages 20 and 55. The average rates for these age groups (See Table 2.2) range between 95 and 99 percent; virtually all married men in this age group are in the labor force. In contrast to the mother's diminished work rates, the fact that children increase the need for family income may raise the male's working time, perhaps by encouraging longer hours of work or a second job for the father. The impact of children on the market activity of older men may be even more pronounced. Bowen and Finegan found that family size was positively correlated with participation rates of married men over age 55.[19] It is possible that these older men are delaying retirement until the last child leaves home or is through school. At that point, income requirements drop and men are freer to retire from the labor force.

Human Capital and Household Time Allocation

During recent years economic analysis has focused new attention on the accumulation of human capital. The analysis holds that people invest in education and training in part because it

19

makes them more productive, and this increased productivity may extend to the production of home or market goods. It would follow that the decision as to whether to allocate more time to market work following the investment in human capital might depend on whether productivity has increased relatively more in market than in nonmarket work.

In the case of married women, the higher their educational attainment, the greater is their level of labor force participation, Table 2.4 indicates that for wives with four years or more of college the work rate is 54.8 percent, while for wives with less than 4 years of high school, the rate is only 32.4 percent. This pattern of increased market activity with increased years of schooling holds when presence and age of children in the family are held constant. When there are no children under age 18 in the family, the LFP is only 28.5 for mothers with less than four years of high school, as contrasted with 65.8 for those with four or more years of college. Thus it appears that education raises market productivity and wages relatively more than it increases home productivity.

The positive correlation between education and participation in market work, drawn from the cross-sectional data, may reflect varying tastes for market work on the part of women. Those wish-

Table 2.4. Labor Force Participation Rates in 1972 of Married Women by Years of Schooling and Age of Children

Presence of young children	Total	Years of school completed		College	
		Less than 4 years of high school	4 years of high school	1 year or more	4 years or more
All wives	41.5	32.4	45.2	48.9	54.8
No children under 18	42.7	28.5	51.2	56.9	65.6
Children 6–17	50.2	44.8	52.2	54.0	60.3
Children under 6	30.1	26.5	30.6	33.1	34.0

SOURCE: U.S. Bureau of Labor Statistics, "Marital and Family Characteristics of Workers, March 1972," *Special Labor Force Report 153* (Washington, D.C.: U.S. Department of Labor, 1973), Table P.

ing to work will invest more in their human capital than those who prefer to remain in the home. On-the-job training in the home may be more useful (certainly it is less costly) than additional years of schooling for wives who do not plan extensive market activity. Over time, improvements in educational attainment may accompany rising family incomes, which in turn allow the purchase of more schooling. A shift in consumer preferences for education also may have stimulated longer schooling.[20]

Average educational attainment for females has risen only slightly during the past 20 years. Median school years for females aged 16 and over has increased from 12.0 to 12.5 since 1952.[21] Yet it should be noted that each successive group of females (or males) is better educated than the one preceding; any cross-sectional survey shows that educational attainment declines with age.[22] Rising levels of education for women therefore seem to explain but little of the initial increase in female labor force activity.

Whether better education and other supply factors stimulate labor force activity, or whether increased demand[23] for female labor induces women to enter the market and then, with jobs available, to invest more in schooling in order to maximize earnings, is not altogether clear. Even if demand conditions have been pulling women into the labor force, the force of demand is more easily met if changing social attitudes, particularly governmental pressure to end discrimination and meet sex quotas, assure new waves of women equality of treatment in the work force. Since it is clear to women that in order to prepare for the better, newly opened positions, they must remain in school longer, the increased enrollment of young women in higher education, particularly in graduate and professional schools, may be largely a response to their current prospects of obtaining more prestigious and higher-paying jobs.

Increased education affects male participation through the same mechanisms—increases in expected wages and access to more appealing and remunerative jobs. In the case of married men of working age, however, there has been much less variation

21

in labor force activity. Apparently these men generally have thought they had little choice as to whether to work. Nevertheless, Bowen and Finegan found a significant positive correlation between educational level and participation rate of males, after controlling for the influence of other variables (age, marital status, color, and other income). In their sample of males in urban areas in 1960, the adjusted participation rates ranged from 90.3 percent for males with less than 4 years of schooling to 99.1 percent for those with 17 or more years of school.

Schooling has a greater impact on market activity of men over age 55. Bowen and Finegan argue that "in making participation decisions older persons as a rule enjoy considerably more freedom from social, economic, and family pressures than do younger persons. Furthermore, employers are extremely cautious about hiring older workers."[24] The lack of formal education may have affected the ability of these older workers to find new jobs after 55, especially in cases where the need arises to change jobs or return to work from a layoff. In general, men with more years of schooling have tended to remain in the labor force, perhaps because of the higher opportunity cost of retiring; in addition, they have been in greater demand than their less-educated peers and more secure in their jobs.

The data in table 2.5 for 1962 and 1973 support these generalizations. In both years there is a strong positive correlation between years of school and participation rate. In the latter year,

Table 2.5. Labor Force Participation Rates for Males Aged 55 to 64, by Educational Level

Years of school completed	March 1962	March 1973	Decrease
Elementary: 8 years or less	83.9	70.9	13.0
High School: 1 to 3 years	89.1	79.3	9.8
4 years	90.6	84.9	5.7
College: 1 to 3 years	89.1	83.9	5.2
4 years or more	93.8	87.0	6.8

SOURCE: Howard Hayghe, "Marital and Family Characteristics of the Labor Force in March 1973," *Monthly Labor Review*, 97 (April 1974), 22.

the participation rate varied from 70.9 percent for males with 8 years of school or less, to 87 percent for those with 4 or more years of college. Observe, too, the relative decline in market activity by years of schooling in the 1962 to 1973 period. The LFP rates of those men with the least education have seen the greatest decline. It is apparent from these data and many studies of problems of older workers that men in their fifties (or even their late forties) who have little education are at a disadvantage in the market and, once they lose their jobs, are unlikely to find new employment. They are also more likely to leave the labor force early in "voluntary" retirement.[25]

Similar patterns of reduced market activity are observed for married women aged 55 to 64. In 1970 the participation rate of all married women with children under 6 years of age was approximately 20 percent. But more than a third of such women with 4 years of high school were in the labor force, whereas among those with 4 or more years of college the participation rate was 41 percent. For women without children under 6, the rate was 28 percent for those with under 8 years of schooling, 39 percent for those with 4 years of high school, and 53 percent for those with 4 or more years of college.[26]

The 1974 figures for all women show a similar pattern. The participation rate for women aged 55 to 64 with 8 years of schooling or less was 33 percent; with 4 years of high school the activity rate was 46.3 percent. Among women with 4 or more years of college, 57.8 percent were in the labor force.[27]

Recent findings require that a note of caution be added to the traditional view on the education-productivity wage relationship. Taubman and Wales state that much of the increase in lifetime earnings attributed to schooling is in fact due to greater ability of college graduates and to the use of the diploma as a screening device by employers.[28] Additionally, Margaret Gordon and the contributors to her volume on higher education and the labor market find that recent graduates were encountering increased difficulties in obtaining the high-status jobs that were available to graduates during the 1950s and 1960s.[29] Such findings indicate that the human capital models which ignore demand factors may produce disproportionately high rates of return to schooling.

What do these trends and comparisons of educational attainment indicate for the analysis of changing life-cycle activities of family members? The length of formal schooling is decided in early life, usually before marriage. This decision has a major influence on when one enters worklife and when he retires. Since one cannot erase years of schooling once they have been completed, nor easily add them in later years, early educational decisions place individuals on different lifetime earnings tracks. Their potential income streams then affect the time allocation of family members by setting the price of their market time throughout subsequent years of life.

During the twentieth century, progressively more people have been able to remain in school for longer periods of time. As a result, a steady influx of more highly educated people into the labor force has tended to place older cohorts, who on the average had less schooling, at a competitive disadvantage. A gradual improvement in the quality of the supply of labor available therefore affects the value of a worker's stock of human capital; greater investments in human capital by younger cohorts lower the comparative advantage and thus the potential earnings of the older groups. With a lowered value placed on their working time, older people choose to leave (or are forced out of) the labor force, which results in some decline in the work rates of older groups even before they reach mandatory retirement age.

3

One-Adult Families and Labor Force Activity

The preceding analysis of changes in the patterns of time allocation within families took account of the declining labor force activity of married men at the beginning and end of worklife and the
increasing market activity of women of all ages. But not all men
and women get married. Nor do all those who marry remain married, nor all who have divorced, remarry. It may well be that a
significant factor affecting the future labor supply will be the current decline in the rate of family formation among younger age
groups and a subsequent increase in the number of one-adult
families. In addition to later marriages, more frequent divorces
swell the size of this group. After reviewing the data on these
trends, attention is directed to the question of the impact these
changes may have on the aggregate labor supply.

Trends in Family Formation

Since the mid-1950s there has been a clear trend on the part of
young adults to delay the first marriage. Table 3.1 shows a rise in
the median age at first marriage of almost a full year in these two
decades. The result is a lengthened period of singleness during
early adulthood. In less than a decade and a half, for example
(from 1960 to 1973), the percentage of women 20 to 24 years of

Table 3.1. Median Age at First Marriage, by Sex, for the United States, 1960 to 1973, and for Conterminous United States, 1890 to 1959

Year	Male*	Female	Year	Male*	Female
1973	23.2	21.0	1956	22.5	20.1
1972	23.3	20.9	1955	22.6	20.2
1971	23.1	20.9	1954	23.0	20.3
1970	23.2	20.8	1953	22.8	20.2
1969	23.2	20.8	1952	23.0	20.2
1968	23.1	20.8	1951	22.9	20.4
1967	23.1	20.6	1950	22.8	20.3
1966	22.8	20.5	1949	22.7	20.3
1965	22.8	20.6			
1964	23.1	20.5	1948	23.3	20.4
1963	22.8	20.5	1947	23.7	20.5
1962	22.7	20.3	1940	24.3	21.5
1961	22.8	20.3	1930	24.3	21.3
1960	22.8	20.3	1920	24.6	21.2
1959	22.5	20.2	1910	25.1	21.6
1958	22.6	20.2	1900	25.9	21.9
1957	22.6	20.3	1890	26.1	22.0

SOURCE: U.S. Bureau of Census, *Current Population Reports*, Series P-20, No. 255, "Marital Status and Living Arrangements: March 1973" (Washington, D.C.: U.S. Government Printing Office, 1973), p. 4.

* Figures for 1947 to 1973 are based on Current Population Survey data supplemented by data from the Department of Defense on marital status by age for men in the Armed Forces. Figures for earlier dates are from decennial censuses.

age who were single rose from 28.4 percent to 38.3 percent. The corresponding increase for males was from 53.1 percent to 57.1 percent. These trends resulted in a population in which one out of every three adults is single, widowed, or divorced.[1]

The increase in the number of young single females may be due in part to the college enrollment of larger proportions of the relevant age groups. Between 1960 and 1972, the number of women attending colleges rose from 1.2 million to 3.5 million. These college women were more likely to remain single while students, and student days more often extended beyond the first college degree. The greater labor force participation and increased career orientation of women may contribute to a later wedding day. In addition, females of the war "baby boom" cohort may be in a marriage squeeze as they reach marriageable age one or two years prior to the males of that cohort. Finally, representatives of the women's movement have argued for a wider range of life styles, and this argument has been persuasive to young women and men.[2]

By contrast, there has been a long-term tendency for a greater percentage of both males and females over age 35 to be married. Table 3.2 shows the continuation of this trend between 1960 and 1970. To date, therefore, delaying the first marriages has not

Table 3.2. Percent Single by Age and Sex, 1973 and 1960

Age	Male			Female		
	1973	1960	Change*	1973	1960	Change*
Total, 14 and over	28.5	25.0	3.5	22.2	19.0	3.2
Under 35	54.5	50.7	3.8	44.2	37.6	6.6
35 and over	6.0	7.8	−1.8	5.1	7.2	−2.1
14 to 17	99.2	99.0	0.2	96.2	94.6	1.6
18	93.3	94.6	−1.3	82.1	75.6	6.5
19	87.2	87.1	0.1	68.8	59.7	9.1
20 to 24	57.1	53.1	4.0	38.3	28.4	9.9
20	76.7	75.8	0.9	59.2	46.0	13.2
21	65.5	63.4	2.1	45.4	34.6	10.8
22	58.9	51.6	7.3	35.7	25.6	10.1
23	46.2	40.5	5.7	28.7	19.4	9.3
24	38.5	33.4	5.1	20.5	15.7	4.8
25 to 29	21.2	20.8	0.4	12.1	10.5	1.6
25	28.3	27.9	0.4	13.7	13.1	0.6
26	26.2	23.5	2.7	13.9	11.4	2.5
27	19.0	19.8	−0.8	12.6	10.2	2.4
28	16.3	17.5	−1.2	10.7	9.2	1.5
29	13.2	16.0	−2.8	9.1	8.7	0.4
30 to 34	9.5	11.9	−2.4	6.6	6.9	−0.3
35 to 39	10.0	8.8	1.2	4.5	6.1	−1.6
40 to 44	5.7	7.3	−1.6	4.8	6.1	−1.3
45 to 54	5.3	7.4	−2.1	3.9	7.0	−3.1
55 to 64	5.5	8.0	−2.5	5.4	8.0	−2.6
65 and over	5.3	7.7	−2.4	6.6	8.5	−1.9

Source: U.S. Bureau of Census, *Current Population Reports*, Series P-20, No. 255, ''Marital Status and Living Arrangements: March 1973,'' (Washington, D.C.: U.S. Government Printing Office, 1973), p. 3.

* Differences shown were derived by using rounded percentages for 1973 and 1960.

affected the incidence of marriage in later life. However, Paul Glick cautions:

A detailed analysis of recent marriage trends has suggested that it is too early to predict with confidence that the recent increase in singleness among the young will lead to an eventual decline in lifetime marriage. However, just as cohorts of young women who have postponed child-bearing for an unusually long time seldom make up for the child deficit as they grow older, so also young people who are delaying marriage may never make up for the marriage deficit later on.[3]

The incidence of divorce has been rising consistently for the last 30 years. The proportion of first marriages of women in their early twenties ending in divorce has tripled, rising from 2.1 percent for that female cohort in 1940 to 6.3 percent for the cohort attaining that age in 1970. Moreover, the percentage of females who are divorced by their earlier thirties has risen from 6.3 percent for that group in 1940 to 15.8 percent for those in 1970.[4]

The increase in the divorce rates of young couples has helped to produce a sharp rise in the lifetime divorce rates of female cohorts. For example, "only 12 percent of the first marriages of women born in 1900 to 1904 are expected to end eventually in divorce, as compared with 18 or 19 percent of those women born in 1920 to 1924 and 25 to 29 percent for women born in 1940 to 1944."[5] Factors stimulating this rise in divorce rates include the liberalization of divorce laws in many states; forced separation of many couples during the late 1960s due to military obligation; lower fertility and hence fewer children to complicate separations; and, possibly, downgrading of marriage as an institution.

The combination of later marriages and higher divorce rates (see figure 3.1) has produced a relative decline in the number of husband-wife families and a significant increase in the number of primary individuals.[6] Table 3.3 shows the change in the composition of households since 1969. Husband-wife families as a proportion of all households declined from 70.9 to 67 percent during this period, with the largest decline occurring for families with heads under 35 years of age. There was a slight increase in the per-

Figure 3.1. Trends in Marriage and Divorce Rates, 1920 to 1970

SOURCE: Paul Glick and Arthur Norton, "Perspectives on the Recent Upturn in Divorce and Remarriage," *Demography*, 10 (August 1973), 303.

NOTE: First marriage rates per 1,000 single women, divorced rates per 1,000 married women, and remarriage rates per 1,000 widowed or divorced women: United States, three-year averages, 1921 to 1971.

centage of female-headed families, while the proportion of primary individuals rose from 18.4 to 21.4 percent of all households. This family realignment meant that there were 2 million more primary individuals in the United States in 1973 than there were in 1970.[7]

Living outside the bonds of marriage does not, of course, preclude motherhood for a woman. In fact, many women living without a spouse bear the responsibility for supporting families with small children. Sar Levitan notes that in 1973 there were 708,000 women without husbands but with preschool-age children. This represented a 60 percent increase over the 450,000 women who were in this situation in 1962.

The number of children affected by marital instability is also increasing. For example, the total number of preschool-age children declined by 2.4 percent between 1970 and 1973, while the number of children in households headed by females increased by

Table 3.3. Households by Type and Age of Head, 1974 and 1969 (numbers in thousands)

Type of household	1974				1969			
	Total	Under 35	35-64	65 and over	Total	Under 35	35-64	65 and over
Total households	69,859	20,188	35,791	13,878	61,805	15,349	34,440	12,014
Percent	100.0	100.0	100.0	100.0	100.0	100.0	100.0	100.0
Primary families	78.6	81.5	85.4	56.8	81.6	88.2	86.6	58.8
Husband-wife	67.0	70.1	73.2	46.5	70.9	79.1	75.7	46.6
Other male head	2.0	1.4	2.3	2.1	2.0	1.2	2.0	2.8
Female	9.6	10.0	9.9	8.2	8.7	7.8	8.9	9.5
Primary individuals	21.4	18.5	14.6	43.2	18.4	11.8	13.4	41.2
Living alone*	19.1	13.6	13.6	41.6	16.7	8.7	12.5	39.1
Male	8.1	11.1	5.8	9.6	6.3	6.6	5.0	9.7
Living alone*	6.8	7.8	5.3	9.2	5.6	4.9	4.6	9.2
Female	13.3	7.4	8.7	33.6	12.1	5.2	8.5	31.5
Living alone*	12.3	5.7	8.3	32.4	11.2	3.8	7.9	30.0

SOURCE: U.S. Bureau of Census, Current Population Reports, Series P-20, No. 276, "Household and Family Characteristics: March 1974," (Washington, D.C.: U.S. Government Printing Office, 1975), p. 2.
* One-person households.

35 percent. Over 2 million children under the age of 6 were living in female-headed families in 1973.[8]

The rise in number of single-parent families has resulted from the increased tendency for first births to occur outside marriage and from the higher divorce rate. The proportion of first births occurring outside marriage has more than doubled since the late 1950s, rising from 5 to 11 percent by 1971. The higher divorce rate meant that in 1970 only about 70 percent of all children under the age of 18 were living with two natural parents who had been married only once. For black children, the figure was a very low 45 percent, but the corresponding figure for white children was also low at 73 percent.[9]

Will this reordering of family structure continue? Although it is difficult to predict trends in human behavior, Jessie Bernard draws the following conclusion:

The statistics may be telling us something important about the future of marriage—what, for example, the easier availability of abortion might be doing to the marriage rate of young women, or what effect the new life styles might be having. The increasing practice of living in households of unrelated individuals . . . suggests that early marriage, and perhaps later marriage as well, is not so attractive to young women as it once was. . . . I do believe that the marriage rate will tend to go down, the slack to be taken up by other forms of relationship.[10]

It may be too early to announce the end of marriage as an institution, however. Glick cautions us against interpreting the trend toward later marriages to imply an increase in the never-married rate of a specific cohort. Additionally, the high divorce rate is accompanied by a high rate of remarriage; currently four out of every five individuals obtaining a divorce will eventually remarry.[11] Thus, the high incidence of divorce for American families may represent a reshuffling of partners and not a decline in the overall marriage rate.

Market Activity of Nonmarried Women

The labor force participation of nonmarried females is influenced by some of the same factors that determine the market

31

activity of married women: age, presence of children, and education.[12] Despite the rapid increase in the labor force participation rates of married women, females living in households that do not include husbands comprise over 40 percent of all women workers. Widows (many of whom are elderly) have relatively low levels of market activity; however, single women and divorced or separated women participate at a greater rate than do married women.

Marital Status	Percent in Labor Force in 1974
Single	57
Married (husband present)	43
Divorced or separated	73
Widowed	25

Between 1940 and 1960 there was a steady increase in the level of market activity of single women over the age of 25. Figure 3.2 shows the upward shifts of the cross-sectional curves for 1940, 1950, and 1960. The 1970 age profile seemed to indicate that this trend had been reversed, at least temporarily. However, the upward movement in labor force activity of single women has reappeared since 1970. As a result of this increase in the rate of market activity (as illustrated in table 3.4) and of the changing marital patterns discussed earlier, single women are becoming an increasing proportion of the work force in the United States. They accounted for 20 percent of the increase in the labor force between 1973 and 1974 and now represent 9 percent of the total labor force.[13]

In recent years, age per se does not appear to have had much of an effect on the rate of market activity of single women during the usual worklife span, ages 20 to 64. The participation rates currently range from 64 percent for the 55 to 64 age group to 82 percent for the cohort aged 25 to 34.[14] If the oldest group is omitted,[15] the range of variation is narrowed to 10 percentage points. Using 1960 census data, Bowen and Finegan reported that their multiple

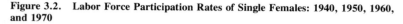

Figure 3.2. **Labor Force Participation Rates of Single Females: 1940, 1950, 1960, and 1970**

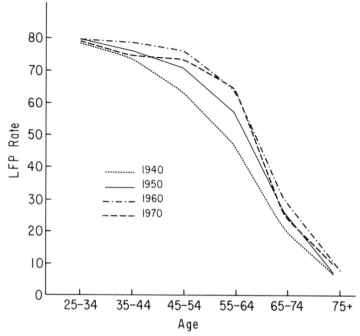

SOURCE: U.S. Bureau of Census, *Census of Population: 1970*, Subject Reports, Final Report PC(2)-6A, "Employment Status and Work Experience" (Washington, D.C.: U.S. Government Printing Office, 1973), Table 5, p. 69.

regression results for single women and Negro single women aged 25 to 54 indicated that "there is no significant association between age and participation. The age-profile for these women over this range apparently can best be described as flat."[16]

Corresponding to our analysis of married women, participation patterns of different cohorts can be drawn (see figure 3.3). Because of the relative stability between age groups within the cross-sectional data and the decline in participation rates between 1960 and 1970, cohort patterns of participation do not differ sig-

Table 3.4. Labor Force Participation Rates of Females by Age Group and Marital Status, 1970 to 1974

	Total*	Under 20*	20–24	25–34	35–44	45–54	55–64	65 and over
Single								
1970	53.0	39.5	71.1	80.7	73.3	72.3	63.7	17.6
1971	52.7	39.6	69.1	77.6	72.8	74.1	65.2	17.4
1972	54.9	41.9	69.9	84.7	71.5	73.0	69.1	19.0
1973	55.8	43.6	70.6	81.7	73.8	73.9	66.5	17.1
1974	57.2	45.6	71.5	81.8	72.5	77.7	64.3	14.6
Widowed, divorced, separated								
1970	39.1	46.5	59.7	65.1	67.9	69.1	54.6	9.9
1971	38.5	44.1	59.9	60.9	67.9	68.4	53.9	8.9
1972	40.1	44.6	57.6	62.1	71.7	69.1	54.9	9.8
1973	39.6	38.1	57.6	64.0	70.7	70.0	52.4	9.1
1974	40.9	46.9	66.1	68.2	69.0	69.6	54.5	8.5

SOURCE: U.S. Department of Labor, *Manpower Report to the President: 1975* (Washington, D.C.: U.S. Government Printing Office, 1975), Table B-2.

* Includes women 16 years and older in March of each year.

Figure 3.3. **Cohort Patterns of Labor Force Activity of Single Women**

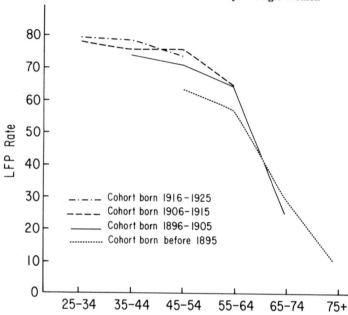

SOURCE: U.S. Bureau of Census, *Census of Population: 1970*, Subject Reports, Final Report PC(2)-6A, "Employment Status and Work Experience" (Washington, D.C.: U.S. Government Printing Office, 1973), Table 5, p. 69.

nificantly from the cross-sectional patterns shown in figure 3.2. There is, however, an upward movement of succeeding cohorts as younger women move through their worklives with greater market attachment.

The age-work profile of previously married women not currently living with their husbands has undergone substantial realignment during the past thirty years. Table 3.5 shows that in 1940 participation peaked around 30 years of age and declined rapidly after age 45. By 1970, the greatest level of market activity

35

Table 3.5. Labor Force Participation Rates of Women Who Are Widowed, Divorced, or Separated, 1940, 1950, 1960, and 1970

Age	1940	1950	1960	1970
14–20	34.6	37.0	35.3	39.8
20–24	57.0	54.3	53.9	61.0
25–29	63.9	59.3	58.2	63.1
30–34	66.6	62.4	62.2	63.9
35–44	61.9	65.7	68.2	68.5
45–54	46.6	56.2	67.3	68.9
55–64	26.8	35.3	47.6	53.8
65–74	8.3	11.8	15.8	15.6
75 and over	2.0	2.2	3.9	4.0

SOURCE: U.S. Bureau of Census, *Census of Population: 1970*, Subject Reports, Final Report PC(2)-6A, "Employment Status and Work Experience" (Washington, D.C.: U.S. Government Printing Office, 1973), Table 5, p. 69.

was by the 45 to 54 age group, and the age profile was almost constant for those women 20 to 54.

With the divorce rate at the highest level in the history of the United States, the number of divorced females in 1974 increased by 353,000. This rise in conjunction with their rising participation rate resulted in an increase of almost 310,000 in the number of divorcees in the labor force, or 13 percent of total increase in the labor force for 1974. As a result, the labor force participation rate of divorced females of all ages rose to 72.9 percent.[17]

For married women at all ages, the presence of young children has a significant (although decreasing) effect of dampening their market activity. But this effect does not hold for women in households without husbands. Table 3.6 shows that such women with children under 6 years of age participated at higher levels than did their counterparts with no small children.

However, the impact of children on participation of nonmarried women is by no means uniform. In 1970, for women under age 25 and over 55, the presence of small children increases participation. However, in the case of women in prime working years (aged 25 to 54), children under 6 retard the level of labor force activity. The relative age distributions of the nonmarried women produce the aggregate result that the participation rate of women

36

Table 3.6. Labor Force Participation of Women Not Classified as "Married, Husband Present," by Age and Presence of Children under Age 6, 1960 and 1970

	1960				1970			
	With own children under age 6		Without own children under age 6		With own children under age 6		Without own children under age 6	
	Number (in thousands)	Percent in labor force	Number (in thousands)	Percent in labor force	Number (in thousands)	Percent in labor force	Number (in thousands)	Percent in labor force
Total	957	46	23,657	41	1586	51	31,842	40
Under 25	284	43	8,776	33	635	49	13,820	36
25–54	659	48	6,045	73	957	52	6,826	73
55 and over	14	36	8,836	26	13	31	11,196	25

SOURCE: U.S. Bureau of Census, *Census of Population: 1970*, Subject Reports PC (2)-5B, "Educational Attainment" (Washington, D.C.: U.S. Government Printing Office, 1973), Table 6; U.S. Bureau of Census, *Census of Population: 1960*, Subject Reports PC (2)-5B, "Educational Attainment" (Washington, D.C.: U.S. Government Printing Office, 1960), Table 5.

with small children is greater than the rate for those without any children under 6. For example, 40 percent of these women with no small children are under the age of 25 and many of them are enrolled in school, thus accounting for their low level of market activity, while 35 percent were over 55 and therefore close to retirement age. On the other hand, over 60 percent of those women with small children were in the prime working ages of between 25 and 54. Thus, the impact of small children on women most likely to be in the labor force was negative; it lowered their labor force participation rate from 73 percent to 52 percent.

The 1974 figures for women who have been married but are not living with spouse show a similar work pattern. These ever-married women with children under 6 participate to a greater degree than those with no children under 18. The age distribution is again the key. For women in the prime working ages, participation is considerably lower with the presence of small children. Mothers aged 25 to 34 with no children under 18 have market activity rates of 83 percent, as compared with 57 percent for women with children under 6. The corresponding figures for women 35 to 44 are 78 percent with no small children and 46 percent with them.[18]

Welfare payments, particularly Aid for Dependent Children (AFDC), provide income to many younger mothers, and the availability of this source of income may induce some women to remain out of the labor force. The number of families receiving AFDC payments tripled between 1961 and 1973, rising to 3 million families that include 7.6 million children. In 1971, AFDC recipients represented more than half of all black female-headed families as well as one in every five white families headed by women. Although there is some trade-off between work and welfare, approximately one-sixth of the women on welfare supplement their transfer income with earnings.[19]

Market Activity of Nonmarried Men

After three decades of decline, single men recently have begun to increase the level at which they participate in market activity.

Figure 3.4 illustrates the decline in labor force participation between 1940 and 1970. This decrease in the rate of participation extended throughout the age spectrum, with the total rate for single males 14 years old and over declining from 62.7 percent in 1940 to 50 percent in 1970. The largest drop was in the rate for men ages 20 to 24, where the level of participation fell from 85 percent in 1940 to 71.6 percent in 1970.[20] This decline corre-

Figure 3.4. **Labor Force Participation Rates of Single Males, 1940, 1950, 1960, and 1970**

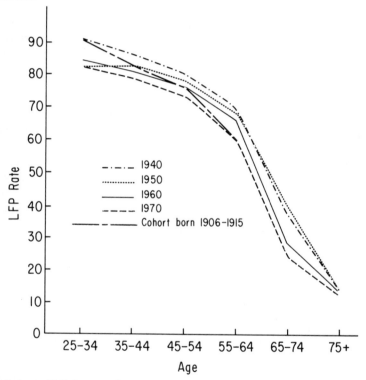

SOURCE: U.S. Bureau of Census, *Census of Population: 1970*, Subject Reports, Final Report PC(2)-6A, "Employment Status and Work Experience" (Washington, D.C.: U.S. Government Printing Office, 1973), Table 5, p. 67.

sponded with the increased enrollment in colleges that occurred during the period.

An analysis of any particular cohort for which we have data would show a steeper decline in market activity with age than do the cross-sectional curves. However, the shape of the age profile of participation would be similar to the cross-sectional patterns (a continuous decline after age 35). For example, the cohort pattern for individuals born between 1906 and 1915 is shown by the short broken lines in figure 3.4. One interpretation of these data could turn on the higher degree of attractiveness of the work-oriented male. If through the marriage search process, women are attracted to men who possess stable work records, the lower participation figures for older males may show only that as a cohort of males age, more and more of those males with strong labor force attachment are acceptable as mates, whereas fewer of the unstable (perhaps for reasons of health, poor education and skills, and the like) are sought by women.

But since 1970, the decline in the rate of participation by single men has been reversed. In the first four years of the decade the labor force participation rate of single men aged 16 years and over rose from 61 to 67 percent.[21] As shown in table 3.7, single men in all age groups under 65 are now more likely to be in the labor force. This pattern of increased participation and continuing demographic trends added two million single males to the labor force during the four years.

Table 3.7.   **Labor Force Participation Rates of Single Men, 1970 to 1974**

Year	Total	Under 20	20–24	25–34	35–44	45–54	55–64	65 and over
1970	60.7	49.0	69.0	86.2	82.3	71.5	60.2	21.0
1971	60.2	47.0	68.5	84.4	79.3	76.8	57.9	21.4
1972	64.5	51.1	73.3	87.5	86.2	81.2	58.6	24.6
1973	66.1	52.6	75.5	87.8	89.4	78.4	66.9	19.6
1974	67.1	54.1	75.5	87.4	87.7	79.6	67.3	15.4

SOURCE: U.S. Department of Labor, *Manpower Report of the President: 1975* (Washington, D.C.: U.S. Government Printing Office, 1975), Table B-2, p. 251.

Population growth among young adults accounted for about half the increase in the number of single labor force participants and a combination of other factors accounted for the rest. For the men, there was the gradual ending of both the military draft and the United States involvement in the Vietnamese War.[22]

The end of the war allowed the return of veterans to the labor force and meant that young men no longer had to be in school to avoid induction into the armed forces. This factor, combined with rising educational costs and a relative shortage of attractive jobs for college graduates, may have dissuaded many young men from enrolling in college. Even though the level of college enrollment continues to rise, the proportion of men under age 25 enrolled in college fell from 46 percent in 1969 to 39 percent in 1972. The corollary of this shift is a greater proportion of young males in the labor force; the participation of men 16 to 24 enrolled in school is 45.2 percent, while the rate for those not enrolled is 92.8 percent.[23]

Teen-age participation rates have followed a similar pattern. The level of market activity for whites aged 16 to 17 declined from 51.2 percent in 1948 to 42.4 percent in 1963. Since the early

Table 3.8. **Labor Force* Status of Once-Married Males Not Currently Living with Spouse, by Decade, 1940 to 1970**

Age	1970	1960	1950	1940
Total	58.4	58.2	60.1	64.3
20–24	83.1	83.6	78.5	83.6
25–29	83.9	82.1	76.0	84.1
30–34	84.7	82.1	78.7	85.1
35–44	84.0	81.7	80.3	84.4
45–54	79.9	79.8	78.2	82.0
55–64	65.3	68.2	69.1	73.6
65–74	23.8	27.5	38.7	38.4
75 and over	8.3	11.1	12.3	12.2

SOURCE: U.S. Bureau of Census, *Census of Population: 1970*, Subject Reports, Final Report PC(2)-6A, "Employment Status and Work Experience" (Washington, D.C.: U.S. Government Printing Office, 1973), Table 5, p. 67.
 * The labor force includes all persons 14 years and older classified in the civilian labor force plus members of the Armed Forces (persons on active duty with the United States Army, Air Force, Navy, Marine Corps, or Coast Guard).

41

1960s, this rate has risen continuously to 53.3 percent in 1974. For whites aged 18 to 19, the decline was from 76.2 percent in 1948 to 65.4 percent in 1967, while the subsequent rise has carried the participation rate up to 73.6 percent. For blacks the declines have been significantly steeper—from 60 to 32 percent for males aged 16 to 17 and from 81 to 59 percent for those 18 to 19—and the upturn in participation did not occur until the 1970s.[24] For black teen-agers, the current staggeringly high unemployment rates have the effect of discouraging many potential participants from entering the labor force,[25] thereby producing misleading unemployment and labor force participation rates for this group.

Table 3.9 shows the reversal of the downward trend in participation. Since 1968 the participation rate for such men 16 years and older has risen over 12 percentage points to 65.9 percent. This upturn has been characterized by some stabilizing in the rate of labor force activity of older nonmarried males and a sharp increase in the participation of those aged 25 to 44.

Widowed men are considerably less likely to be in the labor force than are divorced males. The age distribution of these groups contributes to the relatively low 35 percent participation rate of widowers. Divorced men are more often in the labor force than single men, but there is little difference between market activity of divorced men and that of married men living with

Table 3.9. **Labor Force* Participation Rates of Males Who Are Widowed, Divorced, or Separated, 1968 to 1974**

Age	1968	1969	1970	1971	1972	1973	1974
Total	53.6	54.1	54.2	55.0	62.7	62.5	65.9
20–24	68.4	72.9	73.2	84.6	88.4	90.3	92.1
25–34	81.9	80.7	74.5	83.9	91.5	90.6	93.5
35–44	85.4	82.5	80.6	80.6	91.0	91.0	92.1
45–54	80.7	85.1	83.6	77.8	83.6	86.3	84.3
55–64	64.0	60.1	67.8	63.7	64.4	66.5	65.7
65 and over	14.0	14.9	16.5	13.0	17.0	14.1	15.5

SOURCE: U.S. Department of Labor, *Manpower Report of the President: 1975* (Washington, D.C.: U.S. Government Printing Office, 1975), Table B-2, p. 252.

* Data relate to the civilian population 16 and over; members of the Armed Forces living off post are included in population and labor force figures. All data are for March of the respective year.

spouse—80 percent for the divorced and 84 percent for the married males in 1974.[26]

Marital Status of the Labor Force

The preceding discussion focuses on two interrelated developments which affect the pattern of time allocation among American workers. First, there have been recent changes in family formation—later marriage and increased incidence of divorce—which in conjunction with a continued population growth resulted in an increase (between 1970 and 1974) of almost 5 million people 16 years and over who were not living in husband-wife households. The proportion of people 16 years or over who are single, divorced, widowed, or separated increased from 35.4 to 36.4 percent in this four-year period.

If these current patterns of participation continue, the reordering of life styles might augment the movement toward an increasing proportion of women in the labor force. Such increase would occur because single and divorced women have higher participation rates then married women, while single and divorced males have lower levels of market activity than married men. Thus if the trends in family formation continue, the labor force of the future will have an increase in the proportion of females.

The second significant factor is the recent rise in participation rates of both males and females who are not living with a spouse. This upturn, which was more pronounced for males, meant that the increase in number of nonmarried persons accounted for 35 percent of the labor force growth between 1970 and 1974. During this period, the labor force rose by 7.8 million, while the number of nonmarried adults in the labor force increased by 5.3 million. If, however, the participation rates had remained stable, these groups of single adults would have added only 2.5 million workers to the labor force.

Obviously, such shifts are creating a work force that is comprised of a growing number of persons who have not married or who are currently not living with their spouses. In 1970 this group represented 30.6 percent of the labor force, but by 1974 they made up 33.7 percent of the total. It follows that the labor force of the future will likely have more women—especially in light of the continuing rise in married women's participation rates and the decline in the market activity of married men—and more non-married individuals of both sexes.

The influence of marital status on labor market activity of males calls for closer attention. Part of the explanation for the lower work rates of single men may be the mate-selection process noted earlier. As the Appendix shows, the gain to marriage is greater the more complementary is the time use of the partners. Thus men may seek wives who are committed to being homemakers, and females search for husbands who will provide a steady flow of earnings. Therefore, career-oriented women and men with lower levels of market attachment would remain single because they are perceived to be unsatisfactory partners. A structural shift in society—say, greater career interest on the part of all women or lower fertility rates—would imply that women see smaller gains from marriage, with a resulting increase in one-adult families. Such shifts tend to raise participation rates of women and may lower the rates for men.

The sharper decline in black teen-age participation has been augmented by their relatively greater increase in school enroll-ment compared to whites. Additionally, the current lower par-ticipation by blacks may be a function of the alarmingly high rate of unemployment for black youths which in 1974 approached 40 percent for 16 to 17 year olds and 17 percent for blacks aged 18 to 19.[27] Such rates may discourage youths from seeking em-ployment and thus keep them out of the labor force.

The trends in the participation of men who are divorced, sep-arated, or widowed are similar to that of single males—a gradual decline until the late 1960s, followed by a rapid upturn in market activity. Census data, as shown in table 3.8, indicate that between 1950 and 1970 the participation rate for "other marital status" males (widowed, divorced, or spouse absent) 14 years of age and older declined from 64.3 to 58.4 percent. In contrast with single men, virtually all of the decline for these males was concentrated in the over-55 age group.

Appendix

A Lifetime Utility Function

Development of the concept of a lifetime utility function for a household may help to explain the allocation of each family member's time over the lifespan of the household. Since the family has certain options as to the total amount of market work to supply, the timing of this market work, and the allocation of time to the production and consumption of goods and services in the home—as well as options as to the division of home and market work among the family members—the goal of apportioning family time in such a way as to achieve maximum family utility is somewhat comparable to apportioning income to achieve a similar goal. In practical terms, limited demand for labor may set fairly narrow bounds within which certain workers are able to vary their working time.

Economists traditionally have relied on the concept of utility to specify the levels of satisfaction that an individual attains from allocating his resources between market work and consumption. In the usual analysis, income is derived from work and the return on assets held. The utility attributable to income is the satisfaction attained from the consumption of goods and services purchased with the income. Leisure, too, is purchased with income.

But leisure is a special commodity acquired by taking a reduction in income by abstaining from work.

The utility function would take the following form:

$$U = U(x_1, x_2, \ldots, x_n; L)$$

where the x's represent commodities that are purchased in the market and L is leisure time.[1] The rational individual maximizes his utility in a particular period subject to certain constraints. These constraints include: length of time period, the wage rate and other income, and, of course, his individual perferences. The consumer is said to be at an optimal position when the marginal utility per dollar spent on each commodity, including leisure, is the same.

In recent years, Gary Becker and others have recognized that the allocation of time is more complex than is portrayed by the model of an individual dividing his time between work and leisure.[2] In addition, the consumption of goods requires time, and the amount of time called for varies with the specific commodity. The individual combines units of time and market goods to produce commodities that enter the utility function directly. For example, tennis combines an individual's time with a tennis racket, court, and balls. Such commodities are called z_i's and may be written

$$Z_i = f_i (X_e, T_i)$$

where X_e is a vector of goods purchased in the market and used as inputs in the production of the ith commodity, and T_i is the time input required to produce Z_i.

In this model each individual is both a producing agent and a utility maximizer, combining time inputs and market goods within their productive capacities to produce Z_i's. The optimal mix of Z_i's is defined by the individual's preferences as he attempts to maximize his utility function. Thus, if there are n goods,

$$U = U(Z_i, \ldots, Z_n)$$

and this function should be maximized subject to a goods or income constraint and a time constraint.[3] The implication of this formulation is that the marginal utility derived from consuming good Z_i should be equal to its full price, including foregone earnings.

The utility function discussed above may be generalized to include utility or disutility from market and nonmarket activities. Time spent in producing Z_i's and in market work may provide the individual with utility (either positive or negative) independent of the consumption of the final product. Clearly, some people derive direct satisfaction from their work while others despise their jobs and endure the disutility only because they provide income. This same principle holds in the production of any Z_i. For example, a clean bathroom or a mowed lawn provides the individual with satisfaction, while the act of cleaning or mowing may yield disutility. Thus, our utility function should be written

$$U = U(Z_1, \ldots, Z_n; T_1, \ldots, T_n; T_w)$$

where T_w is time spent in market work.

Turning now to the examination of a single-period household utility function, we note first that a family must decide how to allocate its resources in order to maximize its utility function. The family utility function takes the form

$$U = U(Z_1, \ldots, Z_n; T_{1j}, \ldots, T_{nj}; T_{wj})$$

where T_{ij} represents the time input of the jth family member in producing the ith commodity, and T_{wj} represents the time spent in market work of the jth individual. This utility function allows for the possibility that the same Z_i (say, walking the dog) provides the family with different levels of satisfaction depending on which family member accomplishes the task.

The family needs or desires different commodities in accordance with its preferences, as shown in the utility function. These goods or services, our Z_i's, can be produced by each family member in accordance with his individual productive capacities. The productivity of family members may vary between goods, with one family member being more productive in cooking while another is more adept at cleaning. If there were no independent utility derived from time, the family would arrange for each member to produce those goods in which the individual had a comparative productivity advantage.[4]

The price of the market goods input would be the same to all members of the household. The price of the time input (P_{Tj}) would, however, vary for different individuals. If the market opportunity cost of time is used as a proxy for the value of the time of the individual, then the wage rate (Wj) should equal P_{Tj}. Thus, the higher an individual's wage the more costly it is for him to allocate his time to home production and away from market work.

Wage rates are a measure of a person's market productivity. Therefore, productivity (both market and nonmarket) will determine the allocation of time within the family, with each member producing those goods in which he or she has a comparative productive advantage. For example, assume family member A has a greater market productivity, that is, he can command a higher wage, while individual B is more capable in performing most household tasks. In this case, the optimal allocation of time is clear. The first family member would have a strong attachment to the labor force, while the second individual would perform most household tasks. Only those tasks in which A had a sufficiently large absolute productive advantage to offset his higher time costs would be saved for A. Family member B may engage in some market activity (perhaps part-time or part-year work) but would remain responsible for production of home-produced goods.

The above description sounds familiar. It offers a rationale for the allocation of time within families in the United States during most of this century. As shown in the preceding chapter, husbands have a strong lifetime attachment to the labor force. Married women have a weaker tie to market activity and are more likely to be engaged in part-time work. Even the working wife generally retains the responsibility for washing, cleaning, cooking, and so on, while the husband may do those "male" jobs of mowing the lawn or heavy household repairs.[5]

To explain family behavior over time, a multiperiod utility model is needed. To analyze resource allocation over the life of the family, we employ the following lifetime utility function:

$$U = U(Z_1^0, Z_2^0, \ldots, Z_n^0, Z_1^1, \ldots, Z_n^1, \ldots, Z_1^T, \ldots, Z_n^T)$$

The superscripts on the Z_i's represent the time period during which they will be consumed. Thus, depending on one's time preference, the utility derived from the future consumption of any Z_i may be less than the current consumption of that same good.

In the lifetime model, current decisions can and do influence future constraints and allocations. If, for example, one Z_i represents the production of human capital for a family member, future wage rates and thus time costs of this person depend on the level of production of this Z_i. Alternatively, the timing and level of consumption of goods may affect the future desirability of these goods. For example, five ice cream cones consumed over a week may provide more satisfaction than five devoured in a single hour, or a single eight-hour work shift may be preferable to four two-hour shifts.

With this model in mind, it is instructive to review the changes in labor force composition described in Chapter 2. The concept of a lifetime utility function for a household enables us to examine the impact of changes in attitudes, market conditions, and federal legislation on the arguments within the utility function and on the constraints under which utility is maximized by the family.

Postulate, for example, an expansion in the demand for labor in that sector of the economy dominated by female workers. Will women respond to this increase in demand for "female" jobs, and how will the response affect intrafamily work and leisure decisions affecting other family members? Clearly, the opportunity cost of wives' time will increase in certain occupations as a result of shifts in demand for female labor; a steady rise in wage rates increases the cost of the time input in home production, which has been mainly that of women. Or consider the enforcement of antidiscrimination laws, which has raised the relative wages of females and opened career opportunities for women. Combined with changing attitudes toward work and new life styles, these shifts have given rise to a greater labor force attachment by married women. As a result, the wife may now be adding more to the family's utility by participation in market work than by specializing in home production.

The increased availability of market substitutes for goods and services previously produced and offered in the home also influences the availability of married women for market work. Child care, which is a very time-extensive activity traditionally performed in the home, is now provided through a growing number of day care centers or is aided by some tax concessions to working mothers. Were the movement of publicly supported child care to continue to grow, the substitution of market-produced child care for home-produced care would free additional time for market work. Perhaps even more significant than day care is the growing acceptance of working mothers; earlier criticisms seem to have softened and in some areas disappeared.

Changes in attitudes and preferences as to types of work appropriate for each sex also are evident. The notion that the husband whose wife worked was unable to provide adequate support for his family was prevalent earlier in this century. This view implied that T_{wf} (time spent in market work by the female) produced disutility for the family. As this attitude fades and women become more career-oriented, then T_{wf} is likely to become a positive argument in the utility function. The findings of the Parnes studies support the view that a woman's perception of her husband's attitude toward her working is an important determinant of the amount of time any wife spends in market activities. In their sample of women 30 to 44 years old, married white women whose husbands' attitudes were favorable toward market work were in the labor force nearly four times as long as those who reported unfavorable attitudes. Black women whose husbands expressed a favorable attitude averaged 50 percent more weeks in the labor force.[6]

Moreover, attitudes as to which sex engages in the home production of goods also are changing. Among young couples, particularly, the husband is expected to perform some household tasks. This expectation grows out of a shift in attitudes toward the roles of family members; T_{im}, male time used in the production of certain household services, no longer adds significantly more disutility to the family than having the wife perform the same tasks. Changing relative wage rates should also stimulate shifts in the allocation of family members' time to these household tasks.

Over the life of the family, greater labor force participation by the wife allows a greater level of investment in human capital by the husband, the children, or the wife herself. For young couples, a working wife may provide the income that allows her husband to finish his full-time schooling, or vice versa. In older families, market work by the wife frequently pays for additional education of the children. Usually the married woman's market time has been substituted for the potential market activity of other family members, rather than the reverse. The income provided by the wife's market job thus has allowed a greater accumulation of human capital than otherwise would be possible, and the capital has been invested in the husband and children. This flexibility in time allocation among family members has resulted in a higher family utility than would have been attained otherwise. Market activity by married women may also encourage earlier retirement by their husbands. In this example, the wife may be substituting her market activity for the later work effort of her husband. Thus, married men with working wives should be able to afford earlier retirement because of the increased income provided from the extra earnings.

The forces stimulating individuals to search for mates and marry recently has been the focus of economic analysis. In his initial article on marriage, Becker assumes that marriage occurs if, and only if, both participants assume that they will be better off by the union, that is, if marriage is expected to increase the utility of each individual.[7] Let V_M, V_F represent the levels of utility of an unmarried male and female and V_M^{MF} and V_F^{MF} their corresponding utilities if they are married. Marriage would be preferred by both if and only if

$$V_M^{MF} \geq V_M$$

and

$$V_F^{MF} \geq V_F$$

When does marriage increase the utility of individuals? When male and female time are not perfect substitutes for each other. Then, by combining resources and allowing each partner to specialize in the production of those goods in which he or she has a comparative advantage, total output of goods by the two individuals is increased. The gain to marriage is greater the more complementary are the inputs. Thus, the union of one partner trained exclusively for home work and another trained for market work provides the largest payoff from marriage.

One obvious reason for individuals forming a marital bond is the physical and emotional attraction between the sexes and the desire to raise own children. Living in the same household lowers the costs of frequent contact and establishes a union that allows the rearing of children. This natural law of reproduction that requires a male and a female means that the gain to marriage will be positively related to the importance of having children to each of the potential mates. "Hence, persons desiring relatively few . . . children either marry later, end their marriages earlier, or do both."[8]

4

Worklives in Transition: Some Implications

The preceding record of changes in patterns of labor force participation documents trends that have important implications for manpower behavior in the coming years. Supply conditions alone will not determine conditions in the labor force of the future; institutional factors, particularly those that restrict access to jobs, will also affect the size and mix of the work force. By concentrating on supply transitions, however, the probable magnitude of the necessary adjustments can be anticipated. Following a brief summary of the major questions raised by recent shifts in the age and sex composition of the work force, several are considered in somewhat greater detail, with the primary objective of identifying gaps in the information needed for appraising future developments. Finally, attention is directed to the policy questions raised by such changes in work arrangements.

Changes in the Age and Sex Distribution of Work

In the final quarter of the twentieth century, market work in the United States will come to be divided between men and women much more evenly than in the past. Some decline in the labor force activity of women in certain European countries—a fall to 49 percent from 54 percent in Finland during the past decade and a half and to 19 percent from 24 percent in Italy in the past decade —appears to reflect sectoral shifts in economic activity. As the proportion of all workers engaged in agriculture diminishes, some women (who are counted as work force participants working on farms in the above countries) apparently find entrance to industrial jobs difficult and leave the job market. Among countries that already have moved to the stage in which only a small percentage of workers remain in farming, the female activity rate appears to be increasing. All the factors argue against any downturn in women's labor force activity in this country: entrance of cohorts of women with successively higher levels of education; pressure of legal requirements on hiring and promotion practices; declining fertility; rising money wages for women as well as for men; growing public acceptance of working women, married or single; greater freedom of life style, including nonmarriage. At the same time, the length of time males spend in school and in retirement continues to grow.

The division of home work between the sexes is less predictable. Information on present arrangements for intrafamily sharing of household duties is incomplete, but studies that are available indicate only a limited shift of work from female to male. Traditional patterns of work in the home, we may discover, change less swiftly than those in the marketplace. Certain other offsets are occurring, however: some services formerly performed by the wife—food preparation, the making of clothes, even the care of the elderly—often are handled by others. Household technology has reduced the effort necessary to care for the home. Most important of all, reduced family size and longer life expectancy have combined to free several decades of time for woman's market work or other pursuits.

Appraising the quantity and quality of time free of work during the coming years also is difficult, in part because of the lack of precision in definition and measurement and in part because women, particularly, continue to engage in home work during those hours not occupied by their market jobs. By the conventional measure of time free from the job, leisure has grown significantly for the male during this century, as workweeks dropped from more than sixty to less than forty hours, and vacations, holidays, and a score of nonworking years in youth and old age became available. For women, the record is less clear. Between the early twentieth-century rural setting, where the woman was a farm worker as well as household manager, to the present urban scene encompassing a market job and home management, how have her work and nonwork schedules been affected?

The number of working hours per week may have fallen very little for the woman with a family and a market job. Indeed, the strain of rearing a family in an urban environment surely accounts for women's occasional nostalgia for that earlier era. What has changed is the number of children per family; the number of years a woman spends in child rearing and hence the number of years, given her longer life expectancy, she has to devote to other endeavors; the degree of financial independence working women have (or could have) during the second half of life, as well as during their early years of worklife; a growing public awareness of the work-related needs of women, in particular, but in a broader sense, of the needs of the family in which both parents work outside the home.

The more even distribution of market work between the sexes has been occurring in a time when also there has been a reapportionment of such work over age groups. In the case of age, however, the reallocation has been not toward spreading work more uniformly, but rather toward concentrating work in the middle years, leaving youth free for education and training and old age for retirement. Both nonworking periods call for heavier transfers of income which are of necessity drawn from the earnings of the middle age group; both absorb nonworking time that theoretically might be apportioned over the worklife, quite possibly improving the utility of that free time.

These disadvantages notwithstanding, the simplistic explanation of such changes in the age distribution of work has turned on the basic premise that the changes improve productivity. Longer periods of education are required, it is argued, because the labor market calls for ever-higher levels of technical skill. Investments in human capital have borne sufficiently high rates of return to justify educational expenditures. Moreover, the high unemployment rates among teen-agers seem to bear out the thesis that foregone earnings, which constitute a large share of educational costs, are relatively low in comparison with returns from the investment. Although recent reports have cast some doubt as to the impact of higher education on earnings, the time spent in school has not diminished.

Alternative allocations of working and nonworking time could have been made during the twentieth century, particularly the last three decades. It would have been possible, for example, to reduce workweeks, add vacation time, or even provide worker sabbaticals for education and job retraining, as productivity improved and the size of the labor force grew. Reductions in the workyear could have been accomplished also by greater reliance on part-time work for the growing group of teen-agers, women with young children, and older people.

The number of involuntary part-time workers 16 years and over has increased by over a million since 1966 with the increase almost equally divided between males and females. Much of this increase has been concentrated among young workers aged 16 to 24. The number of workers in this age group who worked part-time because of slack work, job changing during the week, material shortages, or inability to find full-time work has more than doubled in the last decade, rising to a 1974 figure of over one million.[1]

There also has been an increase in the number of voluntary part-time workers. Currently 10.5 million individuals are working less than 35 hours per week by choice. This represents a 50 percent increase over the 1966 total of 7.5 million. Forty-one percent of this total are young workers aged 16 to 24. Of all the males voluntarily working part time, 60 percent are under 25 years of age.[2] The Parnes studies concluded that for individuals aged 14 to 24 the most important single factor influencing hours worked was school enrollment. They found that four-fifths of the students who worked were employed for less than 35 hours a week while only one-fifth of the out-of-school youth worked on a part-time basis.[3] Current economic conditions are surely an important factor accounting for the rise in part-time work. The slack in the economy has displaced young persons previously working full time and also prevented new aspirants from finding full-time work. Additionally, some young workers may have been drawn into the part-time job market in order to augment family income while the primary wage earners were out of work.

The rationale for lengthening the retirement span is somewhat equivocal. Economists have always treated leisure as a superior good, and it must follow from that assumption that the more leisure the better. Additional years free of work would thus add to one's lifetime satisfaction (especially so, when they have been "earned") unless disutility from the cut in real income associated with retirement more than offsets the value of the added free time. But employers' justification for compulsory retirement almost always turns on the need to clear the top-level jobs in order to provide upward mobility for the younger workers, who because of their more recent training and greater energy levels are thought to be more productive. In practical terms, the latter argument is compelling, particularly in view of the higher levels of education and training of succeeding cohorts of job applicants and the higher levels of pay that senior employees have achieved late in worklife.

In summary, the motivation for redividing market work between the sexes is different from that which prompts the reapportionment between age groups, and the impact of the two changes will be different. The relationship of the two changes is complex, but certain ties are clear. For example, the married woman's earnings have enabled families to finance additional years of schooling for their children. Similarly, the family with two salaries over an extended portion of worklife can afford earlier retirement than the one that must rely on a single worker's income, assuming the same wage scale. Thus, the capacity to purchase an increase in free time for the male is enhanced by woman's market work. Intrafamily support of the young adults still in school and of older men in retirement offers the male greater flexibility in scheduling his work; market activity during the woman's middle years substitutes for the male's labor force activity at each end of worklife.

In order to accommodate to the new patterns of work, however (or to initiate changes, if these new workstyles are deemed undesirable), a number of issues need to be addressed. It is important to ask, for example, what long-run effects these changes in the composition of the work force will have on unemployment. Assuming that the unemployment rate continues to be calculated in the same manner, a stronger labor force attachment by women will mean that they will be less likely to withdraw when jobs are hard to find. Rather, they will continue in the work force as males do, even when unemployed. On the other hand, if retirement comes to be "normal" at age 60, older people may classify themselves as retired whether or not they would choose to work. The classification scheme is important, not only for unemployment insurance purposes, but also because the range of social policies designed to deal with different groups may not be consonant with future job and income needs.

Closely related to unemployment is the issue of geographical mobility which also may shift with the greater work activity of married females. When two jobs are involved, movements to new locations are harder to arrange, despite the fact that both husband and wife will have competing job offers to choose among. The manpower implications of joint husband-wife job searches, until recently limited to young professionals with relatively flexible views as to location, will be more pervasive as the woman's career becomes more widely accepted. Each partner is likely to operate under stricter constraints as far as job location is concerned, although each enjoys somewhat greater freedom from the need to maximize individual earnings.

Yet another question posed by the changing division of work has to do with investments in human capital. In the past, educational outlays for women have been lower than those made for men, since the expected return from males' education has been higher. Beyond formal education, moreover, employers' expenditures for training have been heavily concentrated on males, whose market worklives were longer and more productive than those of women. But women's current interest in lifetime careers and their higher market earnings call for larger expenditures on their education. And although some recent studies indicating relatively low rates of returns to males' schooling may call for a re-examination of the comparative yields on investments in education and those in physical plant, it is still true that those moneys being spent for human capital may need to be reapportioned between the sexes.

The relationship between investments in human capital and subsequent worklife suggests further that the emerging age as well as the sex composition of the work force may affect the pattern of investments in education. Since annual earnings generally rise through most of worklife, highest earnings are received near the age of retirement. But if retirement age is lowered, the returns from one's investment will fall. Alternatively, small investments in the retraining of middle-aged and older workers could postpone retirement age, if work capacity were declining. Distinctions between retirements necessitated by poor job performance and those brought about by lack of job opportunity may have to be made more clearly, if the age of exit from work continues to decline.

The size of current investments in education and job training will influence strongly the lifetime earnings of today's students. But incomes also will reflect an increasing preponderance of two-career families. Since the higher the educational level of the woman, the more likely she is to work (and the more likely she is to have a high-income husband), family income could come to be less evenly distributed in the future. The income gap between the college-educated, two-career family and the couple with less education and job training, which is subject to more frequent unemployment and probably has a larger family, seems likely to widen. The gap between retirees and workers also will widen under certain conditions: a continued lengthening of the retirement span, economic growth that raises the incomes of workers but not retirees, and high unemployment during the later years of worklife.

Finally, there well may be a concerted effort to measure and/or reward with money wages the contributions made outside the marketplace, particularly in the home. The pressure for pricing such services continues from various sources: those concerned with providing social benefits such as Social Security to housewives; those who long have protested the failure to include home services in the national income accounts; women who argue for the greater dignity attached to paying jobs. In the ensuing discussion, some of the implications of these issues are discussed.

Some Implications of the New Work Patterns

The public's perception of who constitutes the labor force has tended to lag behind major compositional changes that have occurred during the past two decades. As a result, the kinds of problems that are emerging are not necessarily the ones for which policies and programs are being formulated. Rapid growth in the number of female-headed families, for example, may call for different social policy, particularly during periods of severe unemployment, from that fashioned to deal with joblessness among teen-agers or male heads of households. Early retirement by males, moreover, may conceal a rise in actual unemployment among older men, who would prefer to continue working but find no employment opportunities.

Unemployment: Measured and Real The employment problems of women traditionally have arisen from those factors that worsen the work opportunities of all marginal groups: minorities, older men, teen-agers. Despite women's higher levels of education, employers have continued to hold certain views as to what constitute appropriate jobs for women and to give weight in their hiring to possible labor force discontinuity caused by home and child care demands. Therefore, as shown in table 4.1 and figure

Table 4.1. Postwar Unemployment Rates in the United States, by Sex

Year	Both sexes	Male	Female
1947	3.9	4.0	3.7
1948	3.8	3.6	4.1
1949	5.9	5.9	6.0
1950	5.3	5.1	5.7
1951	3.3	2.8	4.4
1952	3.0	2.8	3.6
1953	2.9	2.8	3.3
1954	5.5	5.3	6.0
1955	4.4	4.2	4.9
1956	4.1	3.8	4.8
1957	4.3	4.1	4.7
1958	6.8	6.8	6.8
1959	5.5	5.3	5.9
1960	5.5	5.4	5.9
1961	6.7	6.4	7.2
1962	5.5	5.2	6.2
1963	5.7	5.2	6.5
1964	5.2	4.6	6.2
1965	4.5	4.0	5.5
1966	3.8	3.2	4.8
1967	3.8	3.1	5.2
1968	3.6	2.9	4.8
1969	3.5	2.8	4.7
1970	4.9	4.4	5.9
1971	5.9	5.3	6.9
1972	5.6	4.9	6.6
1973	4.9	4.1	6.0
1974	5.6	4.8	6.7

SOURCE: U.S. Department of Labor, *Manpower Report of the President: 1974* (Washington, D.C.: U.S. Government Printing Office, 1974), p. 230.

4.1, female unemployment rates have tended to exceed male unemployment rates.

As Bergmann has shown, the inability of women to gain access to a wider range of jobs has meant that they have crowded[4] into those occupations open to them, not only bidding down their wages but also increasing their probability of going jobless. Even within occupations, however, female unemployment continues to exceed that of males. Table 4.2 shows that the woman's rate is one and a half times that of the male in clerical jobs, for example, and more than twice that of men in sales work.

Figure 4.1. Unemployment Rates in the United States, by Sex, 1947 to 1974

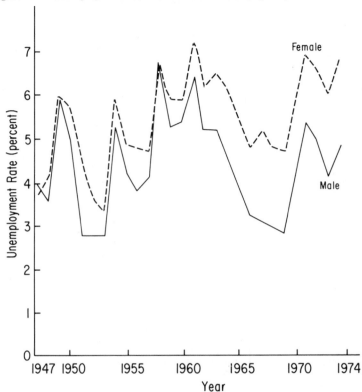

SOURCE: U.S. Department of Labor, *Manpower Report of the President*, *1974* (Washington, D.C.: U.S. Government Printing Office, 1974), p. 230.

During 1974 unemployment rates worsened for men and women, with the female level moving from the 1973 figure of 5.9 to a rate of 7.8 percent late in the year. Among female heads of households—a total of almost 8 million women—the jobless rate was 6.6 percent, as compared with 3.7 percent for male heads. The recession continued to deepen during 1975, with unemployment rates reaching their highest levels since the great depres-

Table 4.2. Unemployment Rates by Occupation and Sex, 1973

	Male	Female
Managerial	1.2	2.5
Professional and Technical	1.7	2.9
Clerical	3.0	4.5
Sales	2.5	5.2
Craft and kindred workers	3.6	5.9
Nonfarm laborers	8.4	9.4
Service workers	5.5	6.0
Transport and communication	4.7	7.8
Primary workers	5.6	3.9
All occupations	3.6	5.0

SOURCE: Calculated from U.S. Department of Labor, *Handbook of Labor Statistics, 1974* in John Leaper, "Female Labor Force Attachment: An Analysis of Unemployment in the United States and Canada," unpublished Ph.D. dissertation, Duke University, 1975.

sion. Economists debate as to whether continued joblessness has the effect of drawing into the work force more secondary workers, including women not attached to the labor force (thereby raising the unemployment rate), or whether discouragement from the job search drives marginal workers out of the market (thus lowering the rate) has persisted.[5]

The importance of this unemployment effect on women's status in the labor force should not be minimized. But as John Leaper points out, the unemployment figure may also be affected by women's participation rates.[6] As female work rates have risen during the last three decades, particularly, their influence on the level of unemployment has been substantial. If women's jobless rates continue to be higher than those of men, yet female participation rises as well, it becomes increasingly important to examine the particular circumstances of female unemployment. Indeed, if the crowding hypothesis is correct, high levels of unemployment will persist until women are integrated into a much wider range of jobs.

To gauge the severity of the job shortage relative to persons who want to work, it is necessary to take into account both those who are defined as unemployed and those who say they want a job, but are not actually looking for work. Nonparticipants in the second category numbered 4.7 million in 1974, more than two-thirds of whom were women. In the fourth quarter of that year 845,000 nonparticipants declared that they were not looking for work, specifically because they were discouraged by the job prospects; again, women made up about two-thirds of this number. Teen-agers and elderly men, along with women, constituted 85 percent of the total.[7] Men of prime working age, by contrast, seldom join this group of nonparticipants.

The number of men in their late fifties and early sixties who classify themselves as nonparticipants, usually as retired rather than unemployed (despite a willingness and capacity to work), is not known. The labor force activity rate for white males aged 55 to 64 has dropped from 89.6 percent in 1948 to a present 78.1 percent; the decline for black males was even sharper: from 88.6 to 70.2 percent. Work rates for 65-and-over white males fell to less than half their 1948 rates (46.5 to 22.5 percent) and faster still for black males (50.3 to 21.7 percent).

Whereas women's participation rates seem likely to grow, those for older men appear slated to continue their decline. Discouragement effects may lower activity rates for both groups somewhat during recessions, but the effect of women's further entrance into the labor force would seem to be much stronger, indicating that high measured rates of unemployment will continue. In addition, the noncounted, discouraged workers, about two-thirds of whom are women, cannot be ignored in policy considerations. Long-range policies aimed at fuller utilization of the labor force will have to focus much more sharply on factors such as job segregation, age discrimination in hiring, and manpower training for women, if either the real or the measured rate of unemployment is to be lowered.

Geographical Mobility of Two-Career Families Early study has shown that the intralabor force mobility of women is lower than that of men.[8] Reasoning that a married woman's location depends more on her husband's place of work than on her own prospects for employment, a recent analysis concluded that "she may find herself unable to migrate in response to her personal opportunities for employment, or she may be forced to migrate when her husband changes jobs."[9] The author argues further that the low mobility rate of the married woman increases both her probability of being unemployed and the chances of her withdrawing from the labor force.

Geographical shifts traditionally have been less acceptable to women than to men. A Bureau of Labor Statistics survey of unemployed persons in 1962 revealed that almost three times as many men as women indicated a willingness to move for a job elsewhere. Married women were particularly reluctant to change locations. Moreover, when women change jobs they stay within the same occupation more often than men. The geographical immobility of women, one author points out, "raises their rate and duration of unemployment relative to those of men."[10]

The lower geographical mobility rates of married women are unlikely to change unless their jobs come to assume greater importance in the family's locational decisions. On the contrary, the wife who works may help to retard her husband's movement to a new job or, at the very least, raise the differential required to induce him to a move. As a result of married women's labor force activity, the willingness of families to move from one place to another thus may be lessened in the future. Some evidence of the growing concern for the wife's career is found in instances of professional couples who choose their locations with due consideration for the availability of two jobs.

Maximizing income and job satisfaction for the two-career family imposes constraints on the movement of both the man and the woman. One effect of such restriction may be the need for some change in corporate policy with respect to the assignment or movement of personnel. Another implication could be a gradual recognition of the importance of the woman's job and indeed a reduction in her unemployment, as she is asked to make fewer job changes in order to relocate with her husband. Whether the reduced mobility of workers will mean a less competitive and a less productive society is a question that needs to be posed. Manpower policy, which has generally espoused high rates of mobility on the theory that geographical moves reduced structural unemployment, will have to take into account any forces which impede such mobility.

Investment in Female Human Capital   The optimal level of investment in the human capital of females is a difficult and underresearched aspect of the human capital literature. The indirect return to females is probably a higher proportion of the total than is the case for males. For example, increased utility may flow to the family because the educated wife is a more efficient homemaker; or early training of children may be enhanced as a result of higher educational attainment of the wife.

Most models of investment in human capital deliberately ignore these returns and concentrate on the increased earnings potential of the female. In these models, since job-related investment by women is directly related to the anticipated length of labor force participation (as in the investment pattern of single women or men), returns to education would not be calculated any differently. However, for women whose worklife cycle is to be interrupted by child rearing, the investment profile will not be monotonically declining. For mothers, the investment in job-related human capital has had two peaks—one before and one after childbearing. The gap may produce some net depreciation as their skills deteriorate.[11]

Human capital models generally make no allowances for barriers to attaining the demanded quantity of investment. Such barriers may have been of major importance in the case of women, especially in on-the-job training and in lack of openings for professional advancement. Investment beyond formal education is dependent in part on the employers' supplying the training. One reason for the scarcity of women in top jobs has been the reluctance of management to provide them with adequate training.

As the labor force attachment of all women increases and

barriers to their entry fall, the investment pattern of females may approach that of males. In light of recent trends, the returns to the investment in human capital of women must receive greater attention. Stefan Hoffer, using four different work histories to estimate the rate of return to years in college for females, finds that these rates for a college degree range from 5.3 percent to 11.7 percent for white women and 8.3 to 17.3 percent for nonwhites. "The study has also shown that when labor force behavior differences between men and women are held constant, all women earned a rate of return on 4 years of college that was larger than that earned by men."[12] It should be noted that similar or even larger rates of return do not imply equal wages; if the opportunity cost of female time is lower during the investment period, equal rates of return would require lower postinvestment wages for women.

The relationship between college education and earnings for males has been analyzed recently by Taubman and Wales,[13] who conclude that the returns attributable to education (as opposed to ability, family background, and other factors) have been overstated. Moreover, the "screening" effect of educational attainment accounts for perhaps half of the returns to a college degree. Increased productivity from education thus is found to be much less important than earlier studies indicated. Although these conclusions were drawn for males only, there is no reason to suppose that market returns on women's work would not reflect the impact of the same variables, including the important factor of screening.

As earlier discussion has indicated, however, rates of returns based only on higher female earnings will understimate the total returns to increases in educational attainment of women. Lee Benham notes that total returns to education for women include higher earnings potential, returns derived through marriage to males of higher earnings streams, and certain benefits that accrue to the family due to the presence of a more highly educated wife. The latter include an increase in the husband's earning potential as a result of his association with a more highly educated wife.[14]

How Will the Distribution of Income Be Affected?   Changes in the allocation of work clearly change the distribution of earnings, which constitute the largest source of money income. The transitions are likely to be quite gradual and the effect of any one shift may be offset by another, leaving the overall distribution between income classes unchanged. In cases where working patterns show that women are doing more market work while older and younger men are less often employed, what are the expected aggregate effects?

The impact of an increase in women's labor force activity on the distribution of earnings depends to a large degree on the characteristics of the women who are entering the market: their educational levels or job skills (as perceived by the employer); marital status and when married; the earnings levels of their husbands. We need to know further whether the factors that in the past have influenced women to join the labor force are the forces now in operation. In particular, analysis of the relationship of age, educational level, marital status, and husband's income to the labor force status of women in recent years may indicate that the traditional correlations no longer apply.

Consider the effect of the husband's income on the wife's labor force activity, for example. Earlier study has shown an inverse correlation between the two; lower-income husbands most often had working wives. The size of husbands' income at which wives' labor supply turned backward has risen through time (from about $2,000 in 1951 to approximately $4,000 in 1960 and to $7,000 to $10,000 in 1974), and there now appears to be a wide range of husband's incomes within which the wives' participation rates vary only slightly. In 1974 the rates were 36 percent for the under $5,000-a-year husbands, rising to over 49 percent for those with between $7,000 and $10,000, and back down to 43 percent for the husbands earning $10,000.

Recent survey data further indicate that the commitment of women to the labor force and to a career of their own is becoming less sensitive to the incomes of their husbands. Of the women aged 30 to 44 in the Parnes sample, 60 percent of the white and

67 percent of the black workers reported that they would continue to work even if they could live comfortably without their earnings. These same women expressed a high degree of job attachment and job satisfaction.[15]

The record of wives' work in relation to their education and own earnings is clear and seemingly in contradiction with the effect of husbands' incomes. Work rates for all women rise with the level of education, from 25 percent for those with 8 years of school to 69 percent for those with 5 or more years of college. For married women with husbands present, the figures are somewhat lower but observe the same relative positions. Throughout the age range of 30 to 60 years, the participation rate for married women with 16 or more years of school remains well above 50 percent, while the rate for those with the least education (less than 8 years) ranges between 25 and 35 percent. On the assumption that women with higher educational attainment marry similarly educated men whose earnings tend to reflect their education, the inverse correlation between wives' participation and husbands' income needs further study.

Given the higher propensity of educated women to take market jobs, it follows that in the future increased market work by women will have its major impact on the incomes of those families already above the median. Lester Thurow points out that in 1969 for the male who earned $6,000 to $7,000 there was a 49 percent probability of his wife working; for the man who earned $25,000 the probability was only 19 percent. He continues:

However, if males who earn high incomes are married to women who could earn high incomes in a perfectly fair and liberated world, then women's liberation will make the distribution of income more unequal. Most of the poor women have already gone to work, and most of the wealthy women have yet to go to work.[16]

His argument is underscored by the record of nonwhite women, whose earnings traditionally have been at the bottom of the scale and who have been in the work force far more frequently than white women of comparable age and with similar child responsibilities. Growth in the earnings of higher-income families resulting from an increase in the work activity of highly educated wives, whose earning capacities, along with those of their husbands, are superior, will, of course, widen family income differentials. Such a trend would be offset in some degree by the increase in numbers of one-adult families, assuming that the increase is evenly distributed among the educational and income levels.

The question of how the distribution of income will be affected by women's greater market work must be considered not only from the point of view of the change in differential between high- and low-income families, but also from the standpoint of how much activity affects men's earnings as compared with those of women. As long as the labor market was balkanized, with the higher-paying positions going to men, the wage differential was maintained even though the proportion of women with their own earnings increased, year by year. But if jobs are integrated, how will male and female earnings be affected?

Estelle James addresses this question in a recent study. National income would rise, she finds, as a result of women's movement into the more productive male jobs, and women as a group would gain wages while men would lose. However, the gains and losses within each sex would vary with race, assuming continued segregation, and with educational level, and the gain to women would far exceed the loss to men. Specifically, the wage loss to men who work year round, full time would range between 8 and 18 percent, with black men, particularly those with a grade-school education, losing most heavily. The wages of white women similarly employed would increase by 50 to 65 percent, the greatest improvements going to college-educated women. Among black women, the greatest gain (56 percent) would accrue to the poorest educated, with the college educated gaining the least. Even with the male's loss in wages, the author notes, most households that include working women would gain because women's increase outweighs men's decrease in earnings. But in households where only the male works, earnings would fall both absolutely and relatively.[17]

The present concentration of two-worker families at the lower-income white and black levels would mean a relative improvement in the economic lot of this group as a result of increased earnings of women, despite the loss to men. However, the further impact of sexual integration of the labor market would be an expansion in the numbers of white upper-class women entering the work force. Thus the share of total income going to white upper-class families would increase.[18]

Longer periods of schooling and retirement would appear at first glance to have the effect of concentrating earnings in the middle years, leaving youth and old age relatively less well off. Incomes earned in early worklife are low, in any case, rising throughout life for most workers with stable jobs. Money incomes then drop sharply for most workers at the time of their retirement and usually remain at the same level except for some instances of cost-of-living supplements. As the retirement phase of life lengthens with the lowering of retirement age, real incomes are eroded by price increases. Moreover, the relative economic position of most elderly persons worsens because those in the work force receive the gains from real growth, whereas the aged's real incomes do not rise with growth.

There can be little doubt that a continuation of the move toward concentrating work in the middle years concentrates earnings as well. The low-income population therefore comes to include an increasing proportion of persons who are not in the work force, at least not full time. Along with the young and the old, a growing percentage of the poor consists of female-headed families, which are increasing in numbers and as a proportion of all families. Since public transfers of income make up the bulk of the income of many female-headed families, as well as that of the elderly, it is clear that the emerging work patterns will call for greater public expenditures in order to spread income more evenly between workers and nonworkers.

Accounting for Nonmarket Work Work done in the home, primarily by women, obviously contributes to the welfare of families and the society. Precise measurement of the value of this work is not possible, however, since no dollar value is imputed to work done outside the marketplace. When a wife shifts from home to market work, it is assumed that her earnings improve the family's economic well-being—indeed, the national income accounts show that total income is increased by the amount of her earnings.

But the family's gains from the additional market work might be revealed as noneconomic, illusory, or negative if more precise estimates were available. The gains are noneconomic if the wife's work makes no net contribution to family income, but nevertheless improves family morale, perhaps by lifting her own. They are illusory if market work is undertaken purely for increased income which is actually wiped out when the costs of foregone services are imputed. Or they are negative when, again, market work is chosen for income alone, but the work-related costs plus the costs of foregone services are clearly higher than earnings.

A family's ability to discern the disadvantages resulting from either the illusory or the negative choice, and then reverse its decision concerning the wife's employment, is complicated by the fact that there is no easy way to estimate the value of the services foregone when the wife goes from home to marketplace. The fact that the market price for domestic services exceeds the value imputed to such services in some homes explains why many wives leave their own home work undone to perform the same services for pay elsewhere. The lower the family income and the greater the need for additional earnings, the lower the imputed value of home work, including child care.

National income accounting compounds the confusion concerning the social value of a movement from home to market work. The fact that labor performed in the home is not rewarded by pay bearing some relation to its value makes it difficult to calculate the net gain from women's growing contribution to market work. Still, the important consideration is not the problem of measurement, but rather the tendency to impute a low market value to services customarily performed in the home. When performed by nonfamily members, therefore, cleaning, laundry, and cooking have brought low wages in the labor market, reflecting the fact that, in most instances, these services bear no price tags. Not only has the buyer been conditioned to view these services as cheap, but the women who do the work are conditioned to think of them in the same way.

The value of work done in the home is further obscured by the fact that free time also lacks a monetary measure. An increase in the amount of market work done by a family is accompanied by some reduction in the amount and quality of home services, but working wives and their families continue to perform those household duties considered essential. An important change occurs in the amount of free time available to the family, however, particularly to the working wife. And although the cost of this reduction is difficult to estimate, some psychic loss is surely suffered with each cut in the amount of time available for rest, recreation, and leisure pursuits.

Confusion arising from the inability to weigh the family's gains from market work by the wife is now upstaged, however, by an insistence that home work be rewarded by wages in the same manner as market work and that women (or men) who perform services in the home receive important fringe benefits such as unemployment compensation and Social Security.[19] Recognizing the difficulties inherent in setting wage rates for housewives, one author recently concluded that "the most realistic proposal is probably to assign half of the employed spouse's income to the household spouse"; any other method of determining the wage is unsatisfactory as long as men are the earners, since members of a family are not likely to accept different living standards.[20] Other bases for setting the home rate of pay have been suggested, including the minimum wage and an opportunity cost calculation.[21]

The manner in which the size of the payment would be determined is perhaps less important than the vigor of the argument itself, particularly when coupled with increased demands for attention to guarantees of income during unemployment and retirement. To make the latter guarantees not on the basis of marital status but rather on the rationale of work, irrespective of where it is done, would call for major changes in social legislation. Even more far-reaching, however, is the continued insistence of active women's groups that home work must be compensated in order to achieve a greater degree of equity between the sexes, to add dignity and to increase the work satisfactions of those who perform home services, and to improve the allocation of women's time between market and nonmarket work by clarifying the costs of their time. These arguments are not likely to disappear as long as a large proportion of women continue to be engaged exclusively in home work.

Summary: Labor Force Composition and Manpower Issues

The preceding review of new working patterns has raised a series of issues of importance to manpower economists: changes in the measured rate of unemployment resulting from increased female work force participation and possible shifts in the concept of unemployment as the worklives of women as well as of younger and older men undergo significant changes; a possible decrease in geographical mobility among two-career families; the impact of higher investments in female human capital and the effect on returns to human capital of reduced worklife for males; the long-run effect of changing work patterns on the distribution of income between income classes and the distribution of earnings between men and women; a persistent argument in favor of the payment of wages and fringes to women for work done in the home. None of these changes will be felt immediately except possibly the rise in unemployment rates as women continue to increase their labor force activity.

In the longer-run context, however, these issues will emerge with greater force. Although earlier projections were reassuring to proponents of the status quo, they were also misleading. Older males' expected decline in labor force participation was much slower than has actually occurred, and projections of women's market rates also have been understated. Indeed, in the case of women's labor force entrance, demographic variables can explain little of the increase during the first half of this century. John Durand reported that in the period from 1920 to 1940 changes in the age composition and family characteristics had a dampening effect on female participation. Even though female participation rose by 3 percentage points, it might have risen by another point if age changes had not restricted its rise.[22] Oppenheimer verifies the net reduction in the 1920 to 1940 period and shows a continuation of the retarding effect of marital status and age composition between 1940 and 1960.[23]

Allowing changes only in age structure and marital status, Table 4.3 shows that the dampening effect of age changes was reduced in the decade of the 1960s. In addition, the recent decline in family formation has meant that changes in marital status are stimulating increased labor force participation by women. Projecting labor patterns to 1985, Table 4.3 shows that age structure changes over the next 10 years will contribute to increased market activity of females by raising their labor force participation rates in the 14 to 64 age range by 0.7 percentage points. If in the fifteen years 1970 to 1985, we allow an increase of, say, 3.5 percentage points in the proportion of the female population aged 20 to 64 that is single (the increase was 1.7 points in the 1960s), raising the single proportion of all women in this age group to 15 percent, the female participation rate rises by an additional percentage point. The rapid rise in the proportion of the labor force made up of divorced women is particularly striking. These demographic changes in the next decade will increase the rate of market activity of women instead of retarding its rise as in the earlier period.

Such a projected increase in the proportion of adults who are single, divorced, or separated can be rationalized in terms of Gary Becker's theory of marriage. As discussed earlier (see Appendix), the propensity to marry depends on the expected gain in utility for each person. This potential gain is a function of the degree of complementarity of the production time of the individuals. The trends analyzed in Chapters 2 and 3 foreshadow a continuing de-

Table 4.3. Effects of Changing Age Composition and Marital Status on the Labor Force Participation Rates of Women, 1940 to 1960, 1960 to 1970, and Projected to 1985

	1940–1960			
	Total*	Married†	Single‡	Other§
1. 1940 (observed)	27.7	14.4	46.5	47.3
2. 1960 (observed)	38.3	32.6	44.2	58.3
3. 1960 (expected, new marriage patterns)	23.2	14.0	38.5	45.6
3′. (expected, old marital patterns)	26.5			
4. Age change (3′ − 1)	−1.2	−0.4	−8.0	−1.7
4′. marital status change (3 − 3′)	3.3			
5. Other factors	15.1	18.6	5.7	12.7

	1960–1970			
	Total	Married	Single	Other
1. 1960 (observed)	38.3	32.6	44.2	58.3
2. 1970 (observed)	45.1	42.6	42.2	61.7
3. 1970 (expected, new marital patterns)	38.7	32.4	43.4	58.1
3′. (expected, old marital patterns)	37.9			
4. Age change (3′ − 1)	−0.4	−0.2	−0.8	−0.2
4′. marital status change (3 − 3′)	0.8			
5. Other factors	6.4	10.2	−1.2	3.6

	1970–1985 Total
1. 1970 (observed)	45.1
2. 1985 (expected, old marital patterns)	45.8
3. Potential age change	0.7

SOURCE: Data for 1940, 1960, and 1970 are calculated from U.S. Bureau of Census, *Census of Population: 1970*, Subject Reports, Final Report PC (2)-6A, "Employment Status and Work Experience" (Washington, D.C.: U.S. Government Printing Office, 1973), Table 5, p. 67. The 1985 figures are derived from population projection series W, without immigration, found in U.S. Bureau of Census, *Current Population Reports* Series P-25, No. 485, "Illustrative Population Projections for the United States: The Demographic Effects of Alternate Paths to Zero Growth" (Washington, D.C.: U.S. Government Printing Office, 1972). The methodology is similar to that employed in Robert Tsuchigane and Norton Dodge, *Economic Discrimination against Women in the United States* (Lexington, Mass.: D. C. Heath, 1974), pp. 70–73, with the allowance for a specific change in marital status.

* Includes participation rate of all women, 14 to 64.

† Includes married women, husband present.

‡ Women who are never married.

§ Includes women who are divorced, separated, widowed, and women whose husbands are not present.

‖ As projected by the Bureau of Census assuming replacement level fertility and no immigration.

crease in the complementarity of time and, therefore, a decline in the incidence of marriage.

Historically, a man who entered into marriage could expect to receive household services and child care from his mate in exchange for providing a source of income for her. Females—often kept out of the market by customs or even legal limitations—traded their skills in home production for claims on their husbands' incomes. The twentieth century has seen not only a reordering of female time toward market work, but also some increased emphasis on the training of women for jobs and an enactment of rules specifying that they be hired.

As a result of improved job opportunities and higher female wage rates, the payoff to marriage may be changing.[24] Males may not expect their mates to spend their entire work effort on the production of home goods to be shared among household members—indeed, the family may be worse off if women do so. Similarly, females need no longer depend solely on men to provide the income to purchase market goods. Becker shows "that a rise in w_f relative to w_m, F's wage rate relative to M's, with the productivity of time in the nonmarket sector held constant, would decrease the gain from marriage if w_f were less than w_m."[25] Shirley Johnson states the proposition more succinctly: "The women's liberation effect appears to have lowered the economic returns to marriage."[26]

If the desire for own children remains as low as current fertility conditions indicate or continues to decline, the returns to marriage will decrease even further. In addition, changing sexual attitudes that permit cohabitation without marriage may force a fundamental shift in the payoff to marriage. Most of the trends examined earlier in this study would seem to lower the gains from marriage and would thus encourage a continuation of the trend toward more one-adult families. Fredericka P. Santos notes that women, particularly, have less to gain from marriage as their own earnings rise: "If a wife earns a substantial proportion of the family income, complementarity between marriage partners tends to be minimal. In such cases, the wife has less to lose financially from termination of the contract than if she were relatively more productive in the home."[27]

Changes in women's worklives have resulted from developments in demand as well as supply. As the service industries have grown, so has the need for skills and styles of the sort traditionally associated with women—white collar, detailed, office skills; teaching; retail store sales; and so on. The married women who joined the labor force a few decades ago were more likely to view themselves as housewives and mothers first and employees second. In this light, acceptance of lower-level jobs was less difficult. The disadvantages of such jobs are clearly evident, however, and women who now recognize their long-term work commitments are beginning to expect a wider range of job choice than has been available to them. Moreover, nonmarried women, who are growing in numbers and as a proportion of the work force, can be expected to wield a significant influence on the quality and level of jobs held by women. Favorable response from the private sector to women's demands for improved access to the better jobs is probably more critical than further government action at present, although continued pressure for compliance with the law is, of course, vital.

Turning to the long-run implications of the decline in the proportion of the male's adult life spent in the labor force, one observes that institutional arrangements largely dictate the patterns of work of older men. Income-maintenance programs for retirement constrain the amount and kind of work men do for pay in their sixties and later. Social Security makes it possible for men to retire at age 65 with full benefits, and industry has institutionalized this as the outside working limit, while frequently encouraging earlier retirement. Government programs do not allow a worker to allocate his years of work during the life span as he sees fit; he cannot take a year off for retraining or recreation and then make it up during his retirement period. Moreover, almost no business firms allow him to take a year's leave of absence from his job. Social Security arranges for a subsidy to nonworkers after age 65 (or 62) but not before. In short, industry practice and government policy bias the individual's time-allocation decision toward working continuously until the mid-sixties, then consuming leisure full time for the rest of life.

It would be possible to conceptualize a plan for removing these constraints. Suppose the government offered to pay benefits equivalent to the current Social Security transfer to those who chose not to work in any given year. Or a more conservative plan might allow a worker to take a year off with benefits after every seven or ten years of working credits, while specifying perhaps a maximum of 45 years of work for each individual. In addition, the worker would need to be guaranteed that he could return to his job the following year.

Having removed the imposed constraints, the effect of possible uses of the year away from market activity could be analyzed. First, the worker could spend his time in leisure. If his work skills did not deteriorate, he could resume work at the end of the year at a rate similar to his old wage. Thus, discounted lifetime income would be lower but lifetime utility presumably would be greater (figure 4.2).

On the other hand, the worker may choose to use this time away from his regular work to invest in himself in the form of additional human capital. He may decide to learn a new skill or

Figure 4.2.   Lifetime Earnings Profile: Year of Leisure

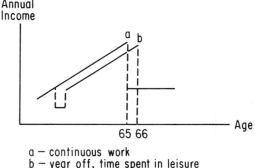

a — continuous work
b — year off, time spent in leisure

80

update or improve his old one. Such activity would increase remaining lifetime income as well as utility (see figure 4.3).

Although there are obvious gains from such reallocation of time, the individual has little freedom to move between work and nonwork pursuits in accordance with his preferences. Concentration of education, work, and leisure in three phases of life creates adjustment problems both at the beginning and end of worklife, as well as maximizing the disutility of work and minimizing the utility of free time. Finally, the packaging of work in the present fashion reduces the individual's flexibility to combine home and market work at any stage of life.

The policies restricting workers' freedom of choice in the timing of work are both public and private. Government income transfer programs, designed to serve the needs of the retired or the unemployed, or, in some instances, youth, have little flexibility. Private industry, too, has largely refrained from introducing variable working arrangements that would accommodate the preferences of employees. Improved scheduling can come about only when industry makes major adjustments by moving to greater use of part-time employment, reducing workweeks, expanding training programs, and so on. Legislation can provide

Figure 4.3. Lifetime Earnings Profile: Year Invested in Human Capital

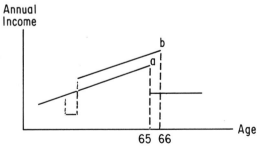

a — continuous work
b — year off, year invested in human capital formation

81

incentives to make such moves (or disincentives not to do so), as in the case of the eight-hour day with premiums for overtime.

In the present era of widespread unemployment, a re-examination of working schedules would seem appropriate. Rearrangements that would allow wider latitude in the choice between home and market work, when each is performed, and how much time is free of work altogether, would help to resolve the issues that are emerging in the wake of a clear preference of women for more market work and a possible desire of men for less. Persistence of large-scale unemployment during the recession of 1974 to 1975 underscores the need for a reconsideration of the interplay between labor force composition and work-nonwork decisions. But the forces which are bringing about the labor force change are of much longer-run significance, and these forces will not be reversed by an upturn in economic activity.

Notes

CHAPTER 1

1. Harold L. Wilensky, "The Uneven Distribution of Leisure," *Social Problems,* 9 (1961), 32–56.
2. Juanita M. Kreps and Joseph J. Spengler, "The Leisure Component of Economic Growth," *Technology and the American Economy,* Appendix Vol. II, National Commission on Technology, Automation, and Economic Progress, 1966).
3. Michael Fogarty, Rhona Rapaport, and Robert Rapaport, *Sex, Career and Family* (London: George Allen and Unwin, 1971), p. 241.
4. See the summary article by Janice N. Hedges and Jeanne K. Barnett, "Working Women and the Division of Household Tasks," *Monthly Labor Review,* 95 (April 1972), 9–14.
5. *Ibid.,* p. 11.
6. Estelle James, "Income and Employment Effects of Women's Liberation," in Cynthia B. Lloyd, *Sex, Discrimination, and the Division of Labor* (New York: Columbia University Press, 1975), pp. 379–391. See also Kathryn Walker and William Gauger, *The Dollar Value of Household Work,* Information Bulletin 60, N.Y. State College of Human Ecology, Ithaca, N.Y., 1973.

CHAPTER 2

1. Steffen B. Lindner, *The Harried Leisure Class* (New York: Columbia University Press, 1970), p. 1.
2. Glen Cain, *Married Women in the Labor Force* (Chicago: University of Chicago Press, 1966), p. 1.

3. U.S. Department of Labor, *Manpower Report of the President: 1975* (Washington, D.C.: U.S. Government Printing Office, 1975).

4. See Cain, *Married Women;* Jacob Mincer, "Labor Force Participation of Married Women," in *Aspects of Economics, A Conference of the Universities,* National Bureau of Economic Research (Princeton: Princeton University Press, 1962); Clarence Long, *The Labor Force under Changing Incomes and Employment,* National Bureau of Economic Research (Princeton: Princeton University Press, 1958); William Bowen and T. Aldrich Finegan, *The Economics of Labor Force Participation* (Princeton: Princeton University Press, 1969); and Valerie Oppenheimer, *The Female Labor Force in the United States* (Berkeley: Institute of International Studies, University of California, 1969); Juanita Kreps, *Sex in the Marketplace: American Women at Work* (Baltimore: Johns Hopkins Press, 1971).

5. This does not mean that the presence of young children in a family does not affect the participation of the mother. It does imply that the market activity of such wives has been rising at a rate that offsets the cross-sectional pattern of decline. See pp. 16–19.

6. U.S. Bureau of Census, *Census of Population: 1970,* Subject Reports, Final Report PC(2)-6A, "Employment Status and Work Experience" (Washington, D.C.: U.S. Government Printing Office, 1973), Table 5, p. 69.

7. Howard Hayghe, "Marital and Family Characteristics of Workers, March 1974," *Monthly Labor Review,* 98 (January 1975), 62.

8. Bowen and Finegan, *Economics of Labor Force Participation,* p. 90. Duran Bell in "Why Participation Rates of Black and White Wives Differ," *Journal of Human Resources,* 9 (Fall 1974), has challenged these results, concluding that the "observed difference in black and white participation rates is real, and it seems to reflect differences in the black responses to the specified variables." Using the 1967 Survey of Economic Opportunity, Bell shows that when black wives are given characteristics that the average white wife possesses the difference in the labor force participation rates of the two groups does not narrow as predicted by Bowen and Finegan, but instead the gap increases.

9. Joseph Gastwirth, "On the Decline of Male Labor Force Participation," *Monthly Labor Review,* 97 (October 1974), 44–46, notes that between the economic peaks of 1955–57 and 1968–69 the labor force participation rate of all males aged 25–54 declined by one percentage point. He estimates that over half of this decline was attributable to more liberal benefits and less stringent requirements for unemployment insurance, welfare payments, and disability benefits. When this effect is combined with the rapid expansion of full-time graduate student enrollment and the change in the definition of labor force status instituted in 1967, Gastwirth "explains" 90 percent of the observed decline.

10. Howard Hayghe, "Marital and Family Characteristics of the Labor Force in March 1973," *Monthly Labor Review,* 96 (April 1974), 21–28. Some part of the decline in the labor force participation of older men likely reflects a decrease in the proportion of males working in agriculture, where they were able to continue working until later in life.

11. U.S. Department of Labor, *The Pre-Retirement Years,* Vol. 3, Manpower Research Monograph 15 (Washington, D.C.: U.S. Government Printing Office, 1972), p. 23.

12. U.S. Bureau of Census, "Employment Status and Work Experience," Table 5.

13. Howard Hayghe, "Marital and Family Characteristics of Workers, March 1974," p. 61.

14. Bowen and Finegan, *Economics of Labor Force Participation,* pp. 51, 290.

15. U.S. Department of Labor, *Dual Careers,* Vol. 2, Manpower Research Monograph No. 21 (Washington, D.C.: U.S. Government Printing Office, 1973), p. 70.

16. U.S. Bureau of Census, "Employment Status and Work Experience," Table 21.

17. U.S. Bureau of the Census, *Census of Population: 1970,* Subject Reports Final Report PC(2)-4A, "Family Composition" (Washington, D.C.: U.S. Government Printing Office, 1973), Table 12.

18. Donald Cymrot and Lucy Mallan, "Wife's Earnings as a Source of Family Income," *Research and Statistics Note Number 10* (Washington, D.C.: U.S. Department of Health, Education, and Welfare, 1974).

19. Bowen and Finegan, *Economics of Labor Force Participation,* p. 296.

20. Cain, *Married Women,* p. 85.

21. U.S. Department of Labor, *Manpower Report of the President: 1974* (Washington, D.C.: U.S. Government Printing Office, 1974), p. 300.

22. *Ibid.,* see p. 302.

23. See Oppenheimer, *Female Labor Force,* Chapter 3.

24. Bowen and Finegan, *Economics of Labor Force Participation,* p. 296.

25. In the case of black men, particularly, who moved from farms to urban settings, reduced labor force activity may also reflect the difference between their former underemployment on the farm and their joblessness and withdrawal from the work force in the city.

26. U.S. Bureau of Census, *Census of Population: 1970,* Subject Reports, Final Report PC(2)-5B, "Educational Attainment" (Washington, D.C.: U.S. Government Printing Office, 1973), Table 6.

27. Beverly McEaddy, "Educational Attainment of Workers, March 1974," *Special Labor Force Report No. 175* (Washington, D.C.: U.S. Bureau of Labor Statistics, 1975), p. 67.

28. Paul Taubman and Terence Wales, *Higher Education and Earnings* (New York: McGraw-Hill, 1974).

29. Margaret Gordon, *Higher Education and the Labor Market* (New York: McGraw-Hill, 1974).

CHAPTER 3

1. See Alix Nelson's review (*New York Times Book Review,* July 6, 1975, p. 4) of several current books dealing with the problems confronting single persons, particularly those with children: Marie Edwards and Eleanor Hoover, *The Challenge of Being Single* (Los Angeles: J. P. Tarche); Persia Woolley, *Creative Survival For Single Mothers* (Millbrae, Cal.: Celestial Arts); Marian Champagne,

Facing Life Alone (New York: Award Books); Michael McFadden, *Bachelor Fatherhood* (New York: Ace Books); Lynn Caine, *Widow* (New York: Bantam Books); Jan Fuller, *Space* (New York: Fawcett).

2. Paul Glick, "A Demographer Looks at American Families," *Journal of Marriage and Family,* 37 (February 1975), 17.

3. *Ibid.,* p. 18.

4. Paul Glick and Arthur Norton, "Perspectives on the Recent Upturn in Divorce and Remarriage," *Demography,* 10 (August 1973), p. 308.

5. *Ibid.,* p. 311.

6. Primary individuals are persons who maintain their own household while living in a house or apartment alone or with persons not related to them.

7. U.S. Bureau of Census, *Current Population Reports,* Series P-20, No. 25, "Marital Status and Living Arrangements: March 1973" (Washington, D.C.: U.S. Government Printing Office, 1973), p. 1.

8. Sar Levitan, *Child Care and ABC's Too* (Baltimore: Johns Hopkins University Press, 1975).

9. U.S. Bureau of Census, *Current Population Reports,* Series P-23, No. 52, "Some Recent Changes in American Families," by Paul Glick (Washington, D.C.: U.S. Government Printing Office, 1975), pp. 3, 9, 10.

10. Jessie Bernard, *The Future of Marriage* (New York: World Publishing, 1972), p. 162.

11. U.S. Bureau of Census, "Some Recent Changes in American Families," p. 2. Glick examines in detail the marital history of the cohort of women born in the 1930s, showing that this group of women set records for early marriage, high birth rates, and low proportion remaining single.

12. We do not discuss the impact of education on the rate of participation of nonmarried females in this chapter. The nature of the effect of increased schooling is similar to that discussed in Chapter 2 for married men and women, that is, higher levels of education raise potential wage rates and induce more individuals to the market. See William Bowen and T. Aldrich Finegan, *The Economics of Labor Force Participation* (Princeton: Princeton University Press, 1969), pp. 254–262; for 1970 data see Bureau of the Census, *Census of Population: 1970,* Subject Reports Final Report PC (2)-5B, "Educational Attainment" (Washington, D.C.: U.S. Government Printing Office, 1973), Table 6.

13. Howard Hayghe, "Marital and Family Characteristics of Workers, March 1974," *Monthly Labor Review,* January 1975, p. 61.

14. U.S. Department of Labor, *Manpower Report to the President: 1975* (Washington, D.C.: U.S. Government Printing Office, 1975), Tables A-1 and B-1.

15. For a recent study of the labor force status of older nonmarried women, see Sally Sherman, "Labor Force Status of Nonmarried Women on the Threshold of Retirement," *Social Security Bulletin,* 37 (September 1974).

16. Bowen and Finegan, *Economics of Labor Force Participation,* p. 262.

17. Hayghe, "Marital and Family Characteristics of Workers, March 1974," p. 61.

18. Howard Hayghe, "Marital and Family Characteristics of the Labor Force, March 1974," U.S. Bureau of Labor Statistics, Special Report No. 173, Table F.

19. Levitan, *Child Care and ABC's Too.*

20. U.S. Bureau of Census, *Census of Population: 1970,* Subject Reports, Final Repōrt PC (2)-6A, "Employment Status and Work Experience," (Washington, D.C.: U.S. Government Printing Office, 1973), Table 5, p. 67.

21. U.S. Department of Labor, *Manpower Report of the President: 1975* (Washington, D.C.: U.S. Government Printing Office, 1975), Table B-2, p. 251.

22. Hayghe, "Marital and Family Characteristics of Workers, March 1974," p. 60.

23. Kopp Michelotti, "Young Workers in School and Out," *Monthly Labor Review,* 96 (September 1973), 12, 13.

24. U.S. Department of Labor, *Manpower Report of the President, 1975,* Table A-4.

25. For a description of the teen-age labor market and the current movement of youth unemployment rates, see Edward Kalachek, *Labor Markets and Unemployment* (Belmont, Cal.: Wadsworth Publishing Company, 1973), or his earlier *The Youth Labor Market,* Policy Papers in Human Resources and Industrial Relations No. 12 (Ann Arbor: Institute of Labor and Industrial Relations, The University of Michigan-Wayne State University, January 1969).

26. Hayghe, "Marital and Family Characteristics of Workers, March 1974," p. 61.

27. *Ibid.,* Table A-19.

APPENDIX

1. Such a utility function is often written $U = U(X, L)$ where X represents the individual's income. This function may then be analyzed using 2-dimension indifference curve analysis. For a detailed analysis, see Belton Fleisher, *Labor Economics* (Englewood Cliffs, N.J.: Prentice-Hall, 1970), pp. 37–50, or Albert Rees, *The Economics of Work and Pay* (New York: Haper and Row, 1973), pp. 22–23.

2. Recent articles include: Gary Becker, "A Theory of the Allocation of Time," *Economic Journal,* 75 (September 1965); Reuben Gronau, "The Intrafamily Allocation of Time: The Value of the Housewives' Time," *American Economic Review,* 63 (September 1973); J. Mincer, "Labor Force Participation of Married Women: A Study of Labor Supply," in *Aspects of Labor Economics,* Universities National Bureau of Economic Research Conference Series 15 (Princeton: Princeton University Press, 1962), pp. 63–105; and G. R. Ghez and G. S. Becker, "The Allocation of Time and Goods over the Life Cycle," Center of Mathematical Studies in Business and Economics, University of Chicago, rep. 7217, April 1972.

3. For a detailed discussion of the properties of this type of production function and the maximization process, see Becker, "A Theory of the Allocation of Time," pp. 495–500.

4. Gronau, "Intrafamily Allocation of Time." Gronau analyzes family production functions and utility maximization in cases when the wife works and also when she does not.

5. Wage rates may also reflect discrimination in hiring and in training. Differences in home productivity may be the result of early training, which is due in part to cultural biases as to who is taught to perform nonmarket production.

6. U.S. Department of Labor, *Dual Careers,* Vol. 1, Manpower Research Monograph No. 21 (Washington, D.C.: U.S. Government Printing Office, 1970), p. 72.

7. Gary Becker, "A Theory of Marriage: Part I," *Journal of Political Economy,* 81 (July/August 1973). Also see G. Becker, "A Theory of Marriage: Part II," *Journal of Political Economy,* 82 (March/April 1974); Alan Friedan, "The U.S. Marriage Market," and T. Dudley Wallace's "Comment," *Journal of Political Economy,* 82 (March/April 1974).

8. Becker, "A Theory of Marriage: Part I," p. 820.

CHAPTER 4

1. U.S. Department of Labor, *Manpower Report of the President: 1975* (Washington, D.C.: U.S. Government Printing Office, 1975), p. 247.

2. *Ibid.,* p. 246.

3. U.S. Department of Labor, *Career Thresholds,* Vol. 1, Manpower Research Monograph No. 16. (Washington, D.C.: U.S. Government Printing Office, 1970), p. 88.

4. Barbara Bergmann, "Curing High Unemployment Rates among Blacks and Women," testimony before the Joint Economic Committee, U.S. Congress, 92d Cong., 2d Sess., October 17, 1972 (Washington, D.C.: U.S. Government Printing Office, 1973). However, several studies have shown that the distribution of the female labor force is not the primary cause of their higher unemployment rate. See Nancy Barrett and Richard Morgenstern, "Why Do Blacks and Women Have High Unemployment Rates?" *Journal of Human Resources,* 9 (Fall 1974), and M. A. Ferber and Helen Lowry, "Women—The New Reserve Army of the Unemployed," presented at the Workshop on Occupational Segregation: Past, Present, and Future, Sponsored by the Committee on the Status of Women in the Economics Profession of the American Economics Association, May 1975. Both of these studies report that if women had the same occupational distribution as men, the female unemployment rate would increase.

5. See W. G. Bowen and T. A. Finegan, "Labor Force Participation and Unemployment," in A. M. Ross, *Employment Policy and the Labor Market* (Berkeley: University of California Press, 1965), pp. 115–161; Jacob Mincer, "Labor Force Participation and Unemployment; A Review of Recent Evidence," in R. A. Gordon and M. S. Gordon, *Prosperity and Unemployment* (New York: Wiley, 1966); Jacob Mincer, "Determining the Number of 'Hidden Unemployed,'" *Monthly Labor Review,* 96 (March 1973), 17–30; Joseph L. Gastwirth, "Estimating the Number of Hidden Unemployed," *Monthly Labor Review,* 96 (March 1973), 17–26; Paul O. Flaim, "Discouraged Workers and Changes in Unemployment," *Monthly Labor Review,* 96 (March 1973), 8–16. For bibliog-

raphy on hidden unemployment and related issues, see the above issue of *Monthly Labor Review*, pp. 31–37.

6. John Leaper, "Female Labor Force Attachment: An Analysis of Unemployment in the United States and Canada," unpublished Ph.D. dissertation, Duke University, 1975.

7. U.S. Department of Labor, *Manpower Report of the President: 1975*, pp. 31–32.

8. Robert L. Bunting, "Labor Mobility: Sex, Race, and Age," *Review of Economics and Statistics*, 42 (May 1960), 229–231.

9. Beth Niemi, "Geographic Immobility and Labor Force Mobility: A Study of Female Unemployment," in Cynthia B. Lloyd, *Sex, Discrimination, and the Division of Labor* (New York: Columbia University Press, 1975), p. 73.

10. *Ibid.*, p. 76.

11. Jacob Mincer and Solomon Polachek, "Family Investment in Human Capital: Earnings of Women," *Journal of Political Economy*, 82 (March/April 1974).

12. Stefan Hoffer, "Private Rates of Return to Higher Education for Women," *Review of Economics and Statistics*, 55 (November 1973), 485. See also Albert Niemi, "Sexist Differences in Returns to Educational Investment," *Quarterly Review of Economics and Business*, 15 (Spring 1975), 20. He finds the rate of return to white females slightly below that to white males for high school and college, but higher for postcollege education. The opposite holds for blacks. Another study by Fred Hines et al., "Social and Private Rates of Returns to Investment in Schooling, by Race-Sex Groups and Regions," *Journal of Human Resources*, 5 (Summer 1970), shows that aggregated over all educational groups, the unadjusted social rates of return were 15.1, 10.2, 6.4, and 10.3 percent for white males, males of other races, white females, and females of other races, respectively.

13. Paul Taubman and Terence Wales, *Higher Education and Earnings* (New York: McGraw-Hill, 1974). See also Richard B. Freeman, "Overinvestment in College Training?" *Journal of Human Resources*, 10 (Summer 1975), 287–311.

14. Lee Benham, "Nonmarket Returns to Women's Investment in Education," in Lloyd, *Sex, Discrimination, and the Division of Labor*, pp. 292–309.

15. U.S. Department of Labor, *Dual Careers*, Vol. 1, Manpower Research Monograph No. 21 (Washington, D.C.: U.S. Government Printing Office, 1970), pp. 207–209.

16. Lester Thurow, "Zero Economic Growth and Income Distribution," in Andrew Weintraub et al., eds., *The Economic Growth Controversy* (White Plains: International Arts and Sciences Press, 1973), p. 146.

17. Estelle James, "Income and Employment Effects of Women's Liberation," in Lloyd, *Sex, Discrimination, and the Division of Labor*, pp. 379–400.

18. *Ibid.*, pp. 387–389.

19. See National Organization of Women, Report of the Task Force on Marriage, Divorce, and Family Relations, October 1972 and NOW Sixth National Conference Resolution, 1973; and other citations in Shirley B. Johnson, "The Impact of Women's Liberation on Marriage, Divorce, and Family Lifestyle," in Lloyd, *Sex, Discrimination, and the Division of Labor*, pp. 401–426.

20. *Ibid.*, p. 419.

21. See Juanita Kreps, *Sex in the Marketplace: American Women at Work* (Baltimore: Johns Hopkins Press, 1971), Chapter 4.

22. John Durand, *The Labor Force in the United States, 1890–1960* (New York: Gordon and Breach Science Publishers, 1968 edition), p. 59.

23. Valerie Oppenheimer, *The Female Labor Force in the United States* (Berkeley: Institute of International Studies, University of California, 1969), p. 26.

24. Samuel Preston and Alan Richards found that "the results of regression analysis support the hypothesis that female economic opportunities have a negative effect on proportions married." See their "The Influence of Women's Work Opportunities on Marriage Rates," *Demography*, 12 (May 1975), 216.

25. Gary Becker, "A Theory of Marriage: Part I," *Journal of Political Economy*, 81 (July/August 1973), 822.

26. Johnson, "Impact of Women's Liberation," p. 423.

27. Fredericka P. Santos, "The Economics of Marital Status," in Lloyd, *Sex, Discrimination, and the Division of Labor*, p. 251.

90

Selected Bibliography

Barrett, Nancy, and Richard Morgenstern. "Why Do Blacks and Women Have High Unemployment Rates?" *Journal of Human Resources,* 9 (Fall 1974).

Becker, Gary. "A Theory of the Allocation of Time," *Economic Journal,* 75 (September 1965).

———. "A Theory of Marriage: Part I," *Journal of Political Economy,* 81 (July/August 1973).

Bell, Duran. "Why Participation Rates of Black and White Wives Differ," *Journal of Human Resources,* 9 (Fall 1974).

Benham, Lee. "Nonmarket Returns to Women's Investment in Education," in Cynthia Lloyd, *Sex, Discrimination and the Division of Labor,* New York: Columbia University Press, 1975.

Bergmann, Barbara. "Curing High Unemployment Rates among Blacks and Women," testimony before the Joint Economic Committee, U.S. Congress, 92nd Congress, 2nd Session, October 17, 1972. Washington, D.C.: U.S. Government Printing Office, 1973.

Bernard, Jessie. *The Future of Marriage.* New York: World Publishing, 1972.

Bowen, William, and T. Aldrich Finegan. *The Economics of Labor Force Participation.* Princeton: Princeton University Press, 1969.

———. "Labor Force Participation and Unemployment," in A. M. Ross, *Employment Policy and the Labor Market.* Berkeley: University of California Press, 1965.

Bunting, Robert. "Labor Mobility: Sex, Race, and Age," *Review of Economics and Statistics,* 42 (May 1960).

Cain, Glen. *Married Women in the Labor Force.* Chicago: University of Chicago Press, 1966.

Cymrot, Donald, and Lucy Mallan. "Wife's Earnings as a Source of Family Income," *Research and Statistical Note Number 10*. Washington, D.C.: U.S. Department of Health, Education, and Welfare, 1974.

Durand, John. *The Labor Force in the United States, 1890–1960*. New York: Gordon and Breach Science Publishers, 1968 edition.

Ferber, M. A., and Helen Lowry. "Women—The New Reserve Army of the Unemployed," presented to Workshop on Occupational Segregation: Past, Present, and Future, Sponsored by the Committee on the Status of Women in the Economics Profession of the American Economic Association, May 21–23, 1975.

Flaim, Paul. "Discouraged Workers and Changes in Unemployment," *Monthly Labor Review*, 96 (March 1973).

Fleisher, Belton. *Labor Economics*. Englewood Cliffs, N.J.: Prentice Hall, 1970.

Fogarty, Michael, Rhona Rapaport, and Robert Rapaport. *Sex, Career and Family*. London: George Allen and Unwin, 1971.

Freeman, Richard B. "Overinvestment in College Training?" *Journal of Human Resources*, 10 (Summer 1975).

Friedan, Alan. "The U.S. Marriage Market," *Journal of Political Economy*, 82 (March/April 1974).

Gaswirth, Joseph. "Estimating the Number of Hidden Unemployed," *Monthly Labor Review*, 96 (March 1973).

———. "On the Decline of Male Labor Force Participation," *Monthly Labor Review*, 97 (October 1974).

Gauger, William, and Kathryn Walker. *The Dollar Value of Household Work*, Information Bulletin 60, New York State College of Human Ecology, Ithaca, N.Y., 1973.

Glick, Paul. "A Demographer Looks at American Families," *Journal of Marriage and Family*, 37 (February 1975).

Glick, Paul, and Arthur Norton. "Perspectives on the Recent Upturn in Divorce and Remarriage," *Demography*, 10 (August 1973).

Gordon, Margaret. *Higher Education and the Labor Market* (New York: McGraw-Hill, 1974).

Gronau, Reuben. "The Intrafamily Allocation of Time: The Value of the Housewives' Time," *American Economic Review*, 63 (September 1973).

Hayghe, Howard. "Marital and Family Characteristics of the Labor Force in March 1973," *Monthly Labor Review*, 97 (April 1974).

———. "Marital and Family Characteristics of Workers, March 1974," *Monthly Labor Review*, 98 (January 1975).

Hedges, Janice, and Jeanne Barnett. "Working Women and the Division of Household Tasks," *Monthly Labor Review*, 95 (April 1972).

Hines, Fred, et al. "Social and Private Rates of Returns to Investment in Schooling, by Race-Sex Groups and Regions," *Journal of Human Resources,* 5 (Summer 1970).

Hoffer, Stefan. "Private Rates of Return to Higher Education for Women," *Review of Economics and Statistics,* 55 (November 1973).

James, Estelle. "Income and Employment Effects of Women's Liberation," in Cynthia Lloyd, *Sex, Discrimination, and the Division of Labor.* New York: Columbia University Press, 1975.

Johnson, Shirley. "The Impact of Women's Liberation on Marriage, Divorce, and Family Lifestyle," in Cynthia Lloyd, *Sex, Discrimination, and the Division of Labor.* New York: Columbia University Press, 1975.

Kalacheck, Edward. *Labor Markets and Unemployment.* Belmont, California: Wadsworth Publishing Company, 1973.

———. *The Youth Labor Market.* Policy Papers in Human Resources and Industrial Relations No. 12. Ann Arbor: Institute of Labor and Industrial Relations, The University of Michigan-Wayne State University, January 1969.

Kreps, Juanita. *Sex in the Marketplace: American Women at Work.* Baltimore: Johns Hopkins Press, 1971.

Kreps, Juanita, and Joseph J. Spengler. "The Leisure Component of Economic Growth," *Technology and the American Economy.* Appendix Vol. II, National Commission on Technology, Automation, and Economic Progress. Washington, D.C.: U.S. Government Printing Office, 1966.

Leaper, John. "Female Labor Force Attachment: An Analysis of Unemployment in the United States and Canada." Unpublished Ph.D. dissertation, Duke University, 1975.

Levitan, Sar. *Child Care and ABC's Too.* Baltimore: Johns Hopkins University Press, 1975.

Lindner, Steffan B. *The Harried Leisure Class.* New York: Columbia University Press, 1970.

Lloyd, Cynthia. *Sex Discrimination, and the Division of Labor.* New York: Columbia University Press, 1975.

Long, Clarence. *The Labor Force under Changing Incomes and Employment.* National Bureau of Economic Research, Princeton: Princeton University Press, 1958.

McEaddy, Beverly. "Educational Attainment of Workers, March 1974," *Special Labor Force Report No. 175.* Washington, D.C.: U.S. Bureau of Labor Statistics, 1975.

Michelotti, Kopp. "Young Workers in School and Out," *Monthly Labor Review,* 96 (September 1973).

Mincer, Jacob. "Determining the Number of 'Hidden Unemployed,'" *Monthly Labor Review,* 96 (March 1973).

———. "Labor Force Participation and Unemployment: A Review of Recent Evidence," in R. A. Gordon and M. S. Gordon, *Prosperity and Unemployment.* New York: Wiley, 1966.

——. "Labor Force Participation of Married Women: A Study of Labor Supply," in *Aspects of Labor Economics,* Universities National Bureau of Economic Research Conference Series 15, Princeton: Princeton University Press, 1962.

Mincer, Jacob, and Solomon Polachek. "Family Investment in Human Capital: Earnings of Women," *Journal of Political Economy,* 82 (March/April 1974).

Niemi, Albert, "Sexist Differences in Returns to Educational Investment," *Quarterly Review of Economics and Business,* 15 (Spring 1975).

Niemi, Beth. "Geographic Immobility and labor Force Mobility: A Study of Female Unemployment," in Cynthia Lloyd, *Sex Discrimination, and the Division of Labor.* New York: Columbia University Press, 1975.

Oppenheimer, Valerie. *The Female Labor Force in the United States.* Berkeley: Institute of International Studies, University of California, 1969.

Preston, Samuel, and Alan Richards. "The Influence of Women's Work Opportunities on Marriage Rates," *Demography,* 12 (May 1975).

Rees, Albert. *The Economics of Work and Pay.* New York: Harper and Row, 1973.

Santos, Fredericka. "The Economics of Marital Status," in Cynthia Lloyd, *Sex, Discrimination, and the Division of Labor.* New York: Columbia University Press, 1975.

Sherman, Sally. "Labor Force Status of Nonmarried Women on the Threshold of Retirement," *Social Security Bulletin,* 37 (September 1974).

Taubman, Paul, and Terence Wales. *Higher Education and Earnings.* New York: McGraw-Hill, 1974.

Thurow, Lester. "Zero Economic Growth and Income Distribution," in Andrew Weintraub, et al., eds., *The Economic Growth Controversy.* White Plains: International Arts and Sciences Press, 1973.

Tsuchigane, Robert, and Norton Dodge. *Economic Discrimination against Women in the United States.* Lexington, Mass.; D. C. Heath, 1974.

U.S. Bureau of Census. *Census of Population: 1970.* Subject Reports, Final Report PC(2)-5B, "Educational Attainment." Washington, D.C.: U.S. Government Printing Office, 1973.

——. *Census of Population: 1970.* Subject Reports, Final Report PC(2)-6A, "Employment Status and Work Experience." Washington, D.C.: U.S. Government Printing Office, 1973.

———. *Census of Population: 1970.* Subject Reports, Final Report PC(2)-4A, "Family Composition." Washington, D.C.: U.S. Government Printing Office, 1973.

———. *Current Population Reports.* Series P-20, No. 276, "Household and Family Characteristics: March 1974." Washington, D.C.: U.S. Government Printing Office, 1975.

———. *Current Population Reports.* Series P-20, No. 255, "Marital Status and Living Arrangements: March 1973." Washington, D.C.: U.S. Government Printing Office, 1973.

———. *Current Population Reports.* Series P-23, No. 52, "Some Recent Changes in American Families," by Paul Glick. Washington, D.C.: U.S. Government Printing Office, 1975.

———. *Current Population Reports.* Series P-25, No. 485, "Illustrative Population Projections for the United States: The Demographic Effects of Alternative Paths to Zero Growth." Washington, D.C.: U.S. Government Printing Office, 1972.

U.S. Bureau of Labor Statistics. *Handbook of Labor Statistics, 1974.* Washington, D.C.: U.S. Government Printing Office, 1974

———. "Marital and Family Characteristics of Workers, March 1972," *Special Labor Force Report Number 153.* Washington, D.C.: U.S. Department of Labor, 1973.

U.S. Department of Labor. *Career Thresholds,* Vol. 1, Manpower Research Monograph No. 16. Washington, D.C.: U.S. Government Printing Office, 1970.

———. *Dual Careers.* Vols. 1 and 2, Manpower Research Monograph No. 21. Washington, D.C.: U.S. Government Printing Office, 1970 and 1974.

———. *Manpower Report of the President: 1974.* Washington, D.C.: U.S. Government Printing Office, 1974.

———. *Manpower Report of the President: 1975.* Washington, D.C.: U.S. Government Printing Office, 1975.

———. *The Pre-Retirement Years.* Vol. 3, Manpower Research Monograph No. 15. Washington, D.C.: U.S. Government Printing Office, 1972.

———. *Years for Decision.* Vols. 1 and 2, Manpower Research Monograph No. 24. Washington, D.C.: U.S. Government Printing Office, 1971 and 1974.

U.S. Department of Labor, Women's Bureau. *Women Workers Today.* Washington, D.C.: U.S. Government Printing Office, 1973.

Wilensky, Harold L. "The Uneven Distribution of Leisure," *Social Problems,* 9 (1961).

Library of Congress Cataloging in Publication Data

Kreps, Juanita Morris.
  Sex, age, and work: The Changing Composition of the Labor Force
  (Policy studies in employment and welfare; no. 23)
  Bibliography: pp. 91–95
    1. Labor and laboring classes—United States. 2. Labor supply—United
States. 3. Time allocation surveys—United States. 4. Women—Employ-
ment—United States. I. Clark, Robert, 1949–   joint author. II. Title.
HD8072.K777   331.1'1'0973   75-34452
ISBN 0–8018–1806–0
ISBN 0–8018–1807–9 pbk.